# Optics in Vision

TO DOROTHY, AS A TOKEN REPARATION

# Optics in Vision

## Foundations of visual optics and associated computations

HENRI OBSTFELD, FBOA, HD, DCLP, FAAO, Optometrist OV (Holland)
*Senior Lecturer in Visual Optics, Department of Applied Optics,
City and East London College*

BUTTERWORTHS
LONDON - BOSTON
Sydney - Wellington - Durban - Toronto

**The Butterworth Group**

| | |
|---|---|
| **United Kingdom** | **Butterworth & Co (Publishers) Ltd** |
| London | 88 Kingsway, WC2B 6AB |
| **Australia** | **Butterworths Pty Ltd** |
| Sydney | 586 Pacific Highway, Chatswood, NSW 2067<br>Also at Melbourne, Brisbane, Adelaide and Perth |
| **Canada** | **Butterworth & Co (Canada) Ltd** |
| Toronto | 2265 Midland Avenue, Scarborough, Ontario, M1P 4S1 |
| **New Zealand** | **Butterworths of New Zealand Ltd** |
| Wellington | T & W Young Building<br>77–85 Customhouse Quay, CPO Box 472 |
| **South Africa** | **Butterworth & Co (South Africa) (Pty) Ltd** |
| Durban | 152-154 Gale Street |
| **USA** | **Butterworth (Publishers) Inc** |
| Boston | 19 Cummings Park, Woburn, Massachusetts 01801 |




First published 1978

ISBN 0 407 00055 0



**British Library Cataloguing in Publication Data**

Obstfeld, Henri
Optics in vision.
1. Eye – Diseases and defects
I. Title
617.7′5 RE48 77–30550

ISBN 0-407-00055-0

Typeset and produced by *Scribe Design* Medway, Kent
Printed litho in Great Britain by W & J Mackay Ltd, Chatham

# Preface

This will not be the first book to be based on the lecture notes of someone who is unhappy with the texts available on his subject. The main purpose of writing it is to provide students of visual optics with a comprehensive account of the geometrical optical aspects of the uncorrected and the corrected ametropic eye, because students seem to be unable to relate the principles of geometrical optics to visual optics.

To my knowledge, no other text shows how virtually all visual optics problems may be solved using 'first principles', nor are alternative methods of solving such problems indicated. From the methods provided the student can select the method which appeals most to him in addition to following the arguments involved in other approaches.

The inclusion of an introductory chapter on geometrical optics proved unavoidable. The reasons for not presenting it in the classical way are the following. Students familiar with geometrical optics will ignore the chapter; thus, a formal derivation of formulae can be omitted. Since the classical approach to optics might deter the less mathematically inclined reader, an attempt was made to make the subject more acceptable by explaining the underlying principles. For a formal approach the reader is referred to, for example, Fincham and Freeman (1974).

In the course of time the origins of visual optics have, to my mind, become somewhat obscured. I, therefore, felt it useful to include some historical details showing that the present approach does not materially differ from earlier ideas. Some clinical aspects of the subject are indicated to round off the introductory chapters.

Having set out the visual optics of the emmetropic eye, the problems of uncorrected spherical ametropia are considered. The principle on which the correction of the ametropic eye is based, is discussed. This is followed by a detailed consideration of the consequences and notions associated with such corrections. The way in which both spectacle magnification and relative spectacle magnification may be employed to determine retinal image sizes, is shown. An equation for relative spectacle magnification involving the distance between the eye's first principal focus and the spectacle point, has been purposely omitted since students do not appear to employ the equation other than with the greatest difficulty. This is probably due to the fact that this distance is not used in any other context.

The senescent changes affecting the eye are enumerated; their visual and optical aspects as well as the effects of their correction, are considered in detail.

Having introduced astigmatism, it is shown that its principle of correction is that of spherical ametropia. I have tried to depict the effects of different retinal image sizes along different meridians in an eye in its simplest form and I felt that photographs would be very helpful.

The application of Newton's equation has been included as an alternative method of solving visual optics problems and as part of the attempt to present a complete text.

Ocular rotations as affected by spectacle corrections, do not appear to have been treated before in the way presented here. It is hoped that the application of 'first principles' will lead to a better understanding.

The calculation of ocular catoptrics and catadioptrics are not usually shown in texts on this subject. For the sake of completeness this and the optics of contact lenses have been included.

For ease of reference a list of symbols and equations has been provided. It will be noticed that the power of a spectacle lens is not indicated by $F_s$ since this is also used to indicate the sagittal power of a lens. Instead, Janet Stone's suggestion to use $F_{sp}$, to avoid confusion, has been adopted. This change as well as a few others, were agreed upon in consultation with A.G. Bennett and R.B. Rabbetts.

The bibliography is limited. I felt that now efficient literature search facilities are widely available, and in view of the avalanche of publications, it would be better to restrict the list. However, a number of references in languages other than English have been included to show that visual optics is a truly internationally discussed subject.

I would like to thank my colleagues for their considerable help in discussing and preparing this work, and for their useful suggestions. My thanks are also due to Michael Hance for the preparation of the photographs in association with Stanley Moss. I am also very grateful to Janet Stone for her help with the chapter on contact lenses.

Over the years I have learned a great deal from discussions with and ideas presented by students. I hope that they will find this book a help.

Any suggestions for improvements will be very welcome.

Henri Obstfeld

# Contents

# Chapter One
# Geometrical Optics

## 1.1 Introduction to basic concepts

The sign convention that will be used throughout this book, is the New Cartesian Sign Convention. This has been used in the teaching of ophthalmic and dispensing optics for many years. The following rules should be observed (*Figure 1.1*):

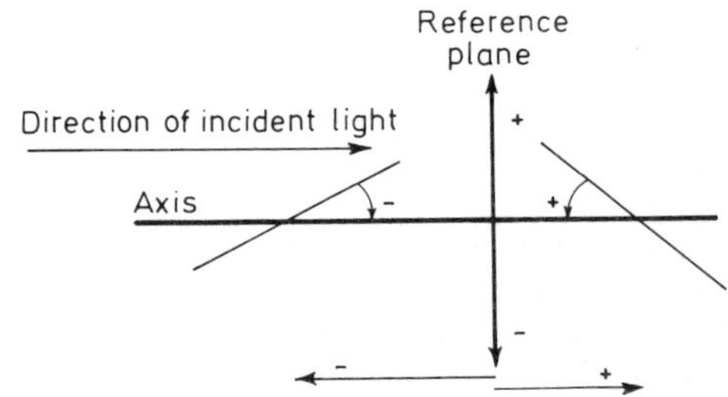


*Figure 1.1 Sign convention*

1. Distances are measured from the reference point or plane (such as a lens, refracting surface or mirror).
2. By convention, light is drawn travelling from left to right. Hence, distances to the right of the reference point are positive; distances to the left are negative (as in the system of Cartesian coordinates). Distances measured upwards from the axis are positive and those measured downwards, negative.
   Acute angles are measured from the ray to the axis. Clockwise angles carry a negative sign and anticlockwise angles carry a positive sign.
3. The 'power' of a lens should be defined in such a way that a converging lens is shown as having positive power and a diverging lens as having negative power (a converging lens has the property of bringing rays of light together so that they cross at a point or focus; a diverging lens has the opposite property).

4. Symbols used to indicate quantities, are printed in italic type; symbols used to indicate geometrical points are printed in roman capital letters. The symbols commonly used include the following:

| | |
|---|---|
| Refractive index | $n$ |
| Object distance | $l$ |
| Image distance | $l'$ |
| First focal length of a lens | $f$ |
| Second focal length of a lens | $f'$ |
| Radius of curvature of a surface | $r$ |
| Object height | $h$ |
| Image height | $h'$ |

Some of the above used terms are self-explanatory; as definition of the others one may say:

Refractive index $n$: the ratio of the speed of light in air (more correctly, the speed of light in vacuum) and the speed of light in a given medium. Since light travels at a speed of about 300 000 km s$^{-1}$, the refractive index of air $n_{air}$ is given by

$$n_{air} = \frac{300\,000 \text{ km s}^{-1}}{300\,000 \text{ km s}^{-1}} = 1.00$$

First principal focal length $f$: the distance from the reference point to the object point for which the image lies at infinity.
Second principal focal length $f'$: the distance from the reference point to the image point for which the object lies at infinity.
(These are sometimes abbreviated to first and second focal length.)
Radius of curvature $r$: the distance from the reference point or surface to its centre of curvature.

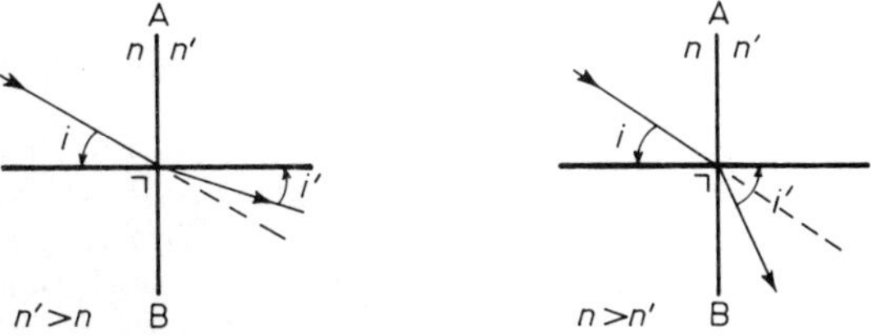


*Figure 1.2 Refraction at a plane surface*

When a ray of light is incident obliquely on a plane surface separating two media having different refractive indices, its direction will change on passing from the first into the second medium. This change of direction at such a surface is known as refraction; the surface is referred to as a refracting surface.

The angle of incidence $i$ on the refracting surface AB (*Figure 1.2*), is

the angle contained between the ray of light at the point of incidence on the refracting surface, and the 'normal' (or perpendicular; but see further) to the surface at that point. The angle of refraction $i'$ is the angle contained between the ray of light in the second medium at the point where the light enters this medium, and the normal to the refracting surface AB at that point.

Snell's law is a statement relating the angles of incidence and of refraction, and the refractive indices of the two media:

$$n \sin i = n' \sin i' \tag{1.1}$$

This law holds good for curved refracting surfaces as well: the point of incidence D (*Figure 1.3*) may be considered as a minute plane surface.

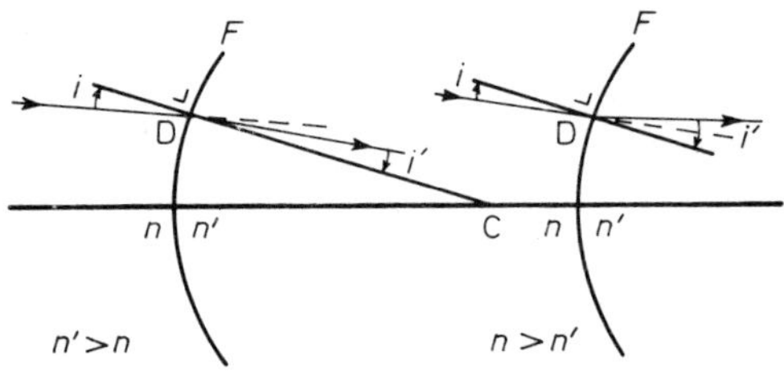


*Figure 1.3 Refraction at a curved surface*

The 'normal' is the line perpendicular to the tangent to the surface at the point of incidence. For a plane surface the tangent will coincide with the surface; for a curved surface it will be perpendicular to the radius of curvature.

The term 'power' indicates the ability of an optical component or system such as a lens or mirror, to change the ray path of incident light: the more a lens changes the ray path, the more powerful the lens is. It will be understood, therefore, that a more powerful lens will have a shorter focal length than a lens of less power.

The curvature of a surface is nowadays measured in a unit known as reciprocal metre, usually written as $m^{-1}$. The reciprocal metre is the reciprocal of the radius of curvature of a surface expressed in metres. Thus, a surface of $r = +1$ m has a curvature $R$ of

$$R = \frac{1}{r} \tag{1.2}$$

$$= \frac{1}{+1}$$

$$= +1 \text{ m}^{-1}$$

From *Figure 1.3* it will be understood that the surface power $F$ is governed by the difference in refractive index of the media on either side of the refracting surface, and the curvature of the surface. (For the derivation of this formula the reader is referred to the textbooks on geometrical optics.) Hence

$$F = (n' - n)R \tag{1.3}$$

$$F = (n' - n)\frac{1}{r} \tag{1.4}$$

The focal power, also designated $F$, of an optical component or system, is measured in dioptres (standard abbreviation D). This unit is defined as the reciprocal of the second focal length $f'$, measured in metres. Since any optical path length is affected by the refractive index in which it is measured, due allowance should be made. Since light travels slower in a medium denser than air, it would travel a longer distance in air than in the denser medium. Consequently, if a distance is measured in a denser medium of refractive index $n'$, the equivalent distance in air or 'reduced distance' would be the factor $n'$ longer than in the denser medium. The dioptre can thus be defined as

$$F = \frac{n'}{f'} \tag{1.5}$$

where $f'$ must be expressed in metres.

Now, if the first focal length were used for this purpose, the power would change sign since the first and second focal lengths are always measured in opposite directions. However, it will be understood that the focal power of a lens cannot change sign. Therefore, the first focal length must be multiplied by the factor (−1) when the power is calculated. Hence

$$F = \frac{-n}{f} = \frac{n'}{f'} \tag{1.6}$$

It follows that

$$n'f = -nf' \tag{1.7}$$

Furthermore

$$f = \frac{-n}{F} \tag{1.8}$$

and

$$f' = \frac{n'}{F} \tag{1.9}$$

The light rays originating from an object or forming an image, are treated in a similar manner. Here we consider the object distance as if it were a focal length. However, this is not called the 'power' of a pencil of light (*Figure 1.4*), but its vergence, which is defined as the reciprocal of the object distance measured in metres with allowance being made

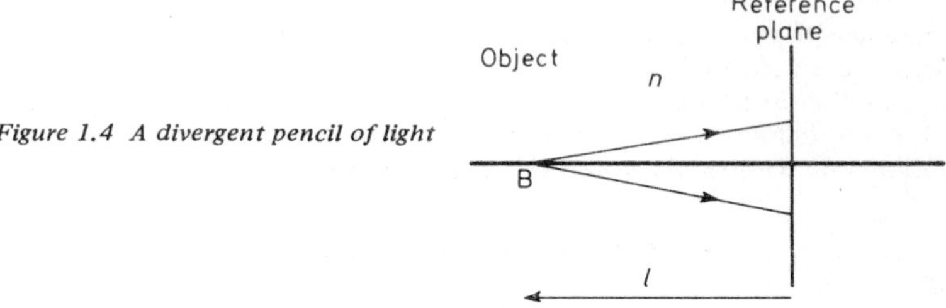


*Figure 1.4 A divergent pencil of light*

for the refractive index in the space between object and reference point (the 'object space'). Calculated in air, it is known as the 'reduced vergence'. Since this definition is basically the same as that of the dioptre, vergences are measured in dioptres. Thus the (reduced) object vergence is given by

$$L = \frac{n}{l} \tag{1.10}$$

Similarly, the (reduced) image vergence is given by (*Figure 1.5*)

$$L' = \frac{n'}{l'} \tag{1.11}$$

where $n'$ is the refractive index of the medium between image and reference point (the 'image space').

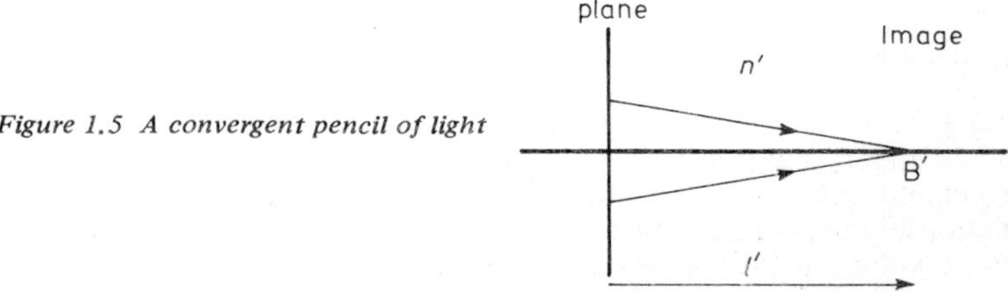


*Figure 1.5 A convergent pencil of light*

In visual optics it is always assumed that calculations refer to that region around the axis of the optical system known as the paraxial region: this is the region around the axis of an optical system containing rays of light of slope angles up to 2° (Emsley, 1956). Within this region we may

make the following approximations without introducing significant errors:

$$\sin w = w$$
$$\cos w = 1$$
$$\tan w = w$$

where $w$ is the angle measured in radians.

Why are we entitled to use this approximation in visual optics? Because the prime area of vision on the retina, the macula, is only about 0.4 mm in diameter. Thus, in a standard reduced emmetropic eye (*see* section 4.4) the macula would subtend an angle of

$$\tan w = \frac{0.4}{22.22} = 0.018\,0$$

which gives an angle of the order of $1°2'$. This is within the $2°$ limit. Another feature of the paraxial region is that rays within it will be incident near the vertex (or point where the optical axis cuts the surface of an optical component) of a spherical surface and, after refraction, they will pass approximately through a point focus (which is a point through which rays of light converge or from which rays of light diverge).

An object point on the axis of an optical system (*Figure 1.6*) and its image point on the axis produced by, for instance, a lens, are known as

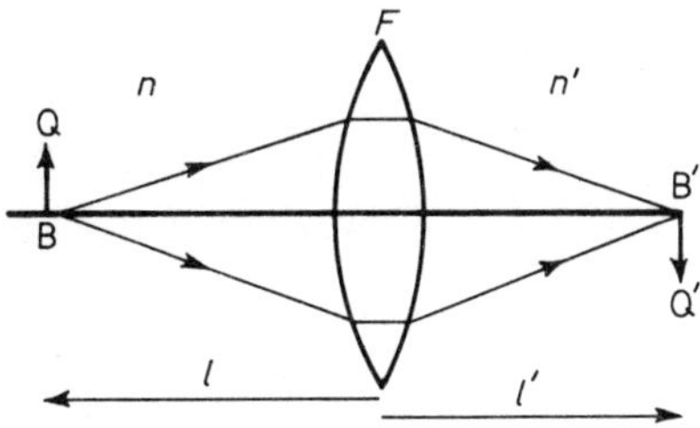


*Figure 1.6 Illustration of the conjugate foci formula*

conjugate points. This concept may be extended to the whole object and its image: object and image planes are conjugate planes. Since the object and image points are foci, the formula which determines the relationship between an object and its image formed by an optical system, is known as the conjugate foci formula:

$$L + F = L' \qquad (1.12)$$

EXAMPLE 1

An object is placed 50 cm in front of a thin converging lens which has a (second) focal length (in air) of +20 cm. Find the position of the image.

NOTE

1. To be able to imagine what the data provided represent, it is best to draw a diagram (*Figure 1.7*). It may now become clear from where the answer to the problem may emerge.

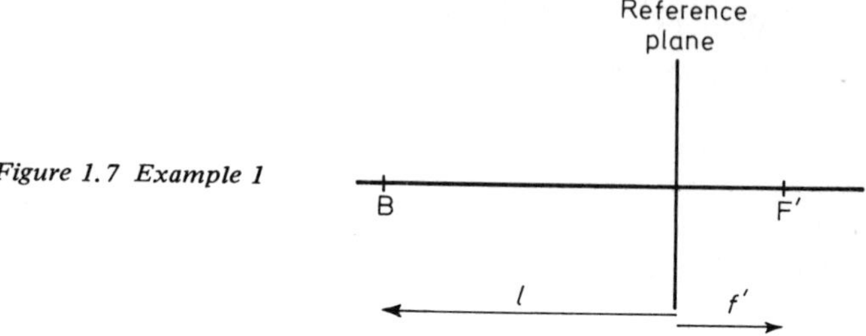


*Figure 1.7 Example 1*

2. Moreover, it may be useful to represent the given data in symbols, thus:

$$l = -50 \text{ cm} = -0.5 \text{ m}$$
$$f' = +20 \text{ cm} = +0.2 \text{ m}$$

3. Often, distances are given in centimetres instead of metres. It is, therefore, common practice to adapt the formulae concerned. The equations cited previously may, where convenient, be written as follows so that distances in centimetres may be substituted:

$$R = \frac{100}{r} \tag{1.2}$$

$$F = (n' - n)\frac{100}{r} \tag{1.4}$$

$$F = \frac{100\,n'}{f'} \tag{1.5}$$

$$F = \frac{-100\,n}{f} = \frac{100\,n'}{f'} \tag{1.6}$$

$$f = \frac{-100\,n}{F} \tag{1.8}$$

$$f' = \frac{100\,n'}{F} \tag{1.9}$$

$$L = \frac{100\,n}{l} \tag{1.10}$$

$$L' = \frac{100\,n'}{l'} \tag{1.11}$$

4. When no specific refractive index is given, it is assumed that the medium concerned is air of refractive index 1.
5. It is advisable to work to two decimal places unless otherwise indicated.
6. In diagrammatic representations positive and negative lenses are represented as shown in *Figure 1.8.* Moreover, an object is always

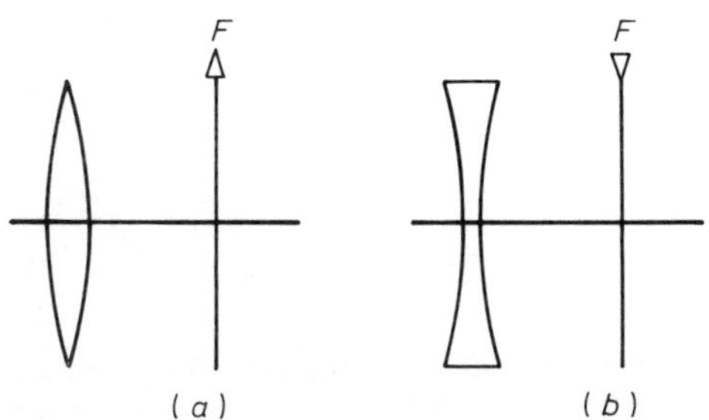


*Figure 1.8 (a) a positive lens at its diagrammatic representation (b) a negative lens and its diagrammatic representation*

indicated by an arrow (*Figure 1.6*). The arrow head is, where necessary, marked by Q and the foot of the arrow by B. Similarly, the image of the object (also an arrow) is marked Q′B′ where necessary. This method of marking cannot be applied to an object or image situated at infinity. In such cases, the angle subtended by the object or image at a given reference point, is used instead. The angles are usually marked $w$ and $w'$.

*Solution 1*

From equation (1.10), the object vergence is

$$L = \frac{100 \times 1}{-50}$$

$$= -2.00 \text{ D}$$

The power of the lens (equation (1.6)) is

$$F = \frac{100 \times 1}{+20}$$

$$= +5.00 \text{ D}$$

From equation (1.12), the image vergence may be calculated, thus:

$$L' = -2.00 + (+5.00)$$
$$= +3.00 \text{ D}$$

Now the image distance may be found by employing equation (1.11):

$$l' = \frac{100 \times 1}{+3.00}$$
$$= +33.33 \text{ cm}$$

which means that the image is formed 33.33 cm to the right of the reference point (the converging lens).

When optical ray paths are constructed graphically, three rules (proof of which will be found in textbooks of geometrical optics) may be applied, namely:

1. A ray of light incident on a positive lens parallel to its optical axis, is refracted and passes through the second principal focus F′ (*Figure 1.9*), or in the case of a negative lens, emerges as if originating from the second principal focus F′ (*Figure 1.10*).

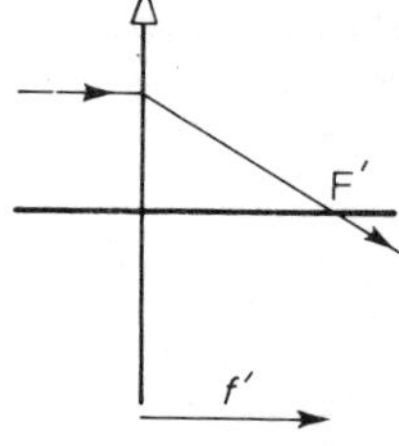


*Figure 1.9 The path of a ray incident parallel to the optical axis of a positive lens*

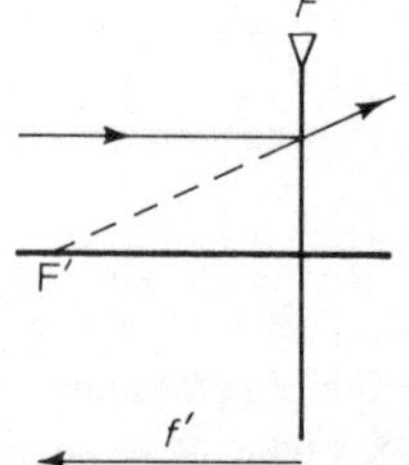


*Figure 1.10 The path of a ray incident parallel to the optical axis of a negative lens*

2. A ray of light passing through the first principal focus F of a positive lens, runs parallel to the optical axis after refraction (*Figure 1.11*), or in the case of a negative lens, a ray of light directed towards the first principal focus F emerges parallel to the optical axis after refraction (*Figure 1.12*).

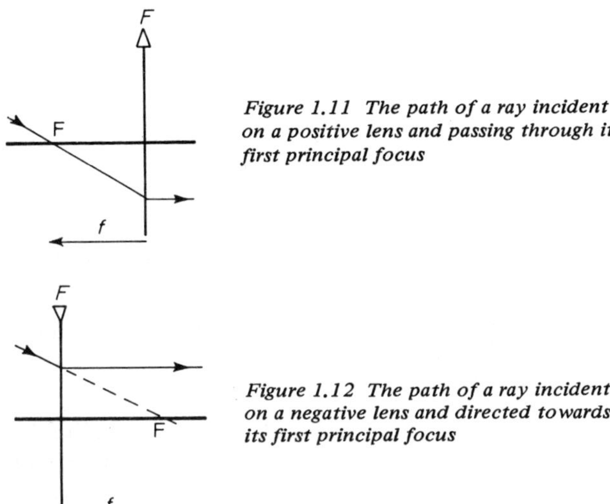


*Figure 1.11 The path of a ray incident on a positive lens and passing through its first principal focus*

*Figure 1.12 The path of a ray incident on a negative lens and directed towards its first principal focus*

3. A ray of light passing through the optical centre of a thin lens, emerges undeviated and undisplaced (*Figure 1.13*) when the same refractive index is present on both sides.

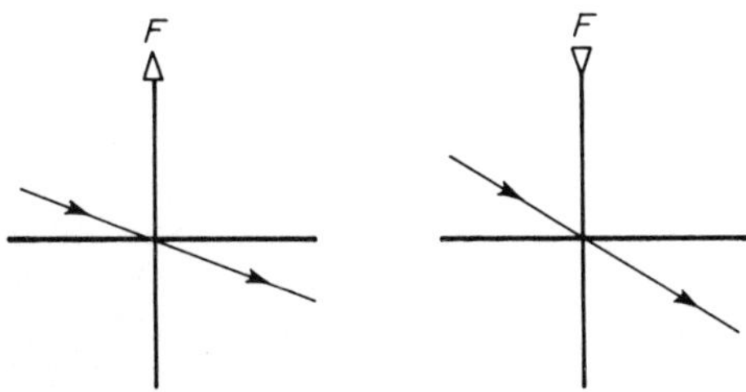


*Figure 1.13 A ray incident on the optical centre of a thin lens, emerges undeviated*

These principles may also be applied when using auxiliary axes (that is, any axis making an angle with the main axis).

Returning to Example 1, we see that the position of the image may now be constructed (*Figure 1.14*), but only after the object point has been extended into an object BQ, since in most cases an optical ray

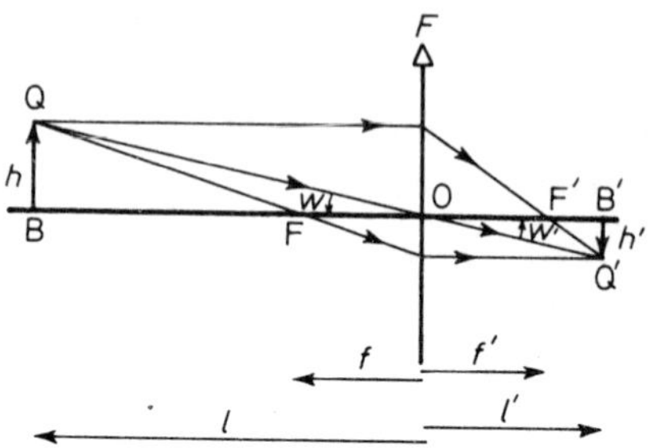


*Figure 1.14 The optical construction of the image position and size*

path cannot be constructed from a point object on the axis of a lens or optical system. The object has a size $h$. *Figure 1.14* shows that only two of the three constructional rays are required to construct the position and size ($h'$) of the image.

From this example further information may be obtained, namely the magnification of the system. Two types of magnification should be distinguished: angular magnification $M$ and lateral or transverse magnification $m$. Angular magnification may be defined as the ratio of the angle subtended by the image of an object, to the angle subtended by the object, both measured at specific reference points. This type will be referred to when discussing spectacle magnification (section 10.1). The angles should be measured in radians.

Lateral or transverse magnification is the ratio of the linear image size $h'$ to the linear object size $h$, or

$$m = \frac{h'}{h} \tag{1.13}$$

Now in *Figure 1.14* it will be seen from ΔΔBQO and B′Q′O that the ratio of image size and object size can also be expressed in terms of image distance and object distance, but only if these distances are measured in media having the same refractive index. Hence:

$$m = \frac{l'}{l} \tag{1.14}$$

Since it is often found to be more convenient to use vergences where the refractive indices are automatically taken into consideration, instead

of distances, equation (1.13) may be modified using Snell's law (equation (1.1)), as follows:

$$\begin{aligned} m &= \frac{l' \tan w'}{l \tan w} \\ &= \frac{l' n}{l n'} \\ &= \frac{n}{l} \times \frac{l'}{n'} \\ &= L \times \frac{1}{L'} \\ &= \frac{L}{L'} \end{aligned} \tag{1.15}$$

A negative value of $m$ indicates that the image is inverted with respect to the object. *Figure 1.14* shows also that the image size is given by

$$h' = -f' \tan w' = -f' \tan w \tag{1.16}$$

This subject leads to the consideration of another well-known formula, namely Newton's relation, which employs the extra-focal distances $x$ and $x'$:

$x$ = the distance between the first principal focus F and the object
$x'$ = the distance between the second principal focus F′ and the image

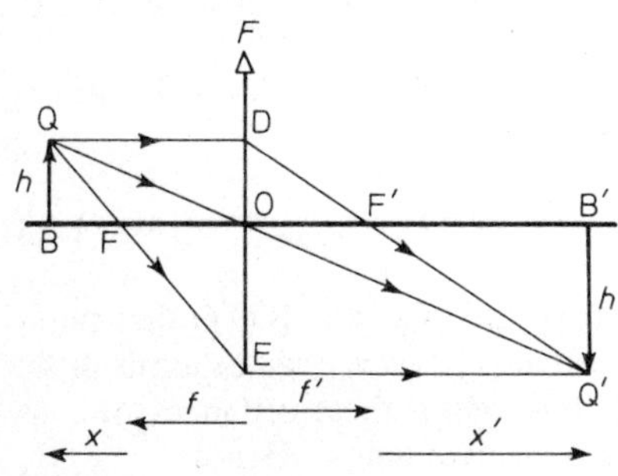


*Figure 1.15 Newton's equation*

From *Figure 1.15* it will be seen that

$$m = \frac{h'}{h}$$

From the diagram ($\Delta\Delta F'Q'B'$ and $F'DO$)

$$\frac{-h'}{h} = \frac{x'}{f'}$$

or

$$m = \frac{-x'}{f'}$$

Similarly ($\Delta\Delta FEO$ and $FQB$),

$$\frac{-h'}{h} = \frac{-f}{-x}$$

or

$$m = -\frac{f}{x}$$

and so

$$-\frac{x'}{f'} = -\frac{f}{x}$$

where the negative sign indicates that the image is inverted with respect to the object.

Finally,

$$ff' = xx' \tag{1.17}$$

which is known as Newton's relation or Newton's equation.

## 1.2 Lens systems

In (visual) optics one often has to consider an optical system consisting of more than one lens. An example is a spectacle lens in combination with an eye. A simple lens system consisting of two thin lenses in air will be considered below.

EXAMPLE 2

A distant object subtends an angle $w$ at the first lens of a lens system consisting of two thin lenses in air. The power of the first lens $F_1 = +15.00$ D and that of the second lens $F_2 = +16.66$ D. The

distance $d$ between the two lenses is 3 cm. Find the position of the image formed by the system.

*Solution 2*

*Step 1.* Draw a diagram representing the above data (*Figure 1.16*).

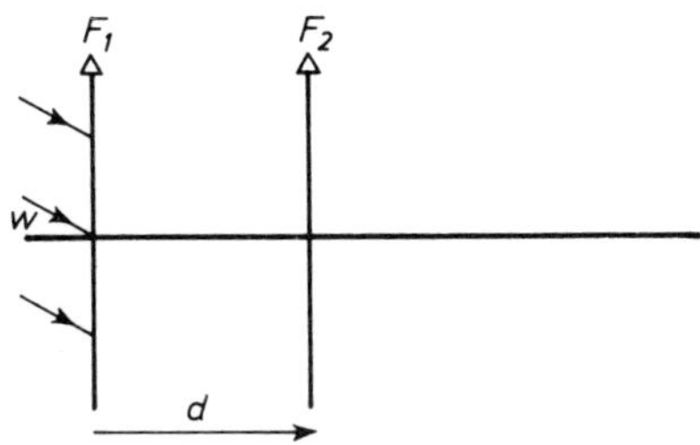


***Figure 1.16** Example 2; a system consisting of two thin lenses in air*

*Step 2.* From the data provided the following information may be calculated:

The second focal length of the first lens (equation (1.9))

$$f_1' = \frac{100 \times 1}{+15.00} = +6.67 \text{ cm}$$

The second focal length of the second lens

$$f_2' = \frac{100 \times 1}{+16.66} = +6.00 \text{ cm}$$

and these focal lengths together with the positions of the principal focal points $F_1'$ and $F_2'$ may now be shown in the diagram (*Figure 1.17*).

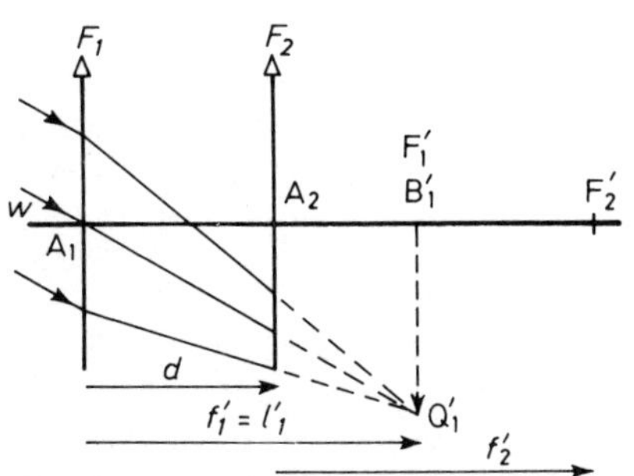


***Figure 1.17** Example 2; the virtual image formed by the first lens of the system*

*Step 3.* Since a distant object is under consideration, the object distance $l = \infty$. Because this is the object distance with respect to the first lens of the system, it will be denoted $l_1$. The object vergence with respect to the first lens, is (equation (1.10))

$$L_1 = \frac{1}{\infty} = 0 \text{ D}$$

*Step 4.* Applying the conjugate foci formula with the appropriate suffixes, the image vergence with respect to the first lens may be calculated (equation (1.12))

$$\begin{aligned} L_1' &= L_1 + F_1 \\ &= 0 + 15.00 \\ &= +15.00 \text{ D} \end{aligned}$$

The image distance is (equation (1.11))

$$l_1' = \frac{100 \times 1}{+15.00} = +6.67 \text{ cm}$$

NOTE

It now appears that the image coincides with the position of the second principal focus of the first lens $F_1'$. In view of the object distance ($l_1 = \infty$) and its associated vergence ($L_1 = 0$ D), this could have been anticipated. Returning to the diagram (*Figure 1.17*), it will be noticed that the image is formed *behind* the second lens. It will be clear that under normal circumstances this image cannot be formed since the second lens is situated in the path of the image forming rays and affects their course. However, for optical construction purposes one imagines that the second lens has no effect, and one completes the diagram with this difference that any ray continuing beyond the second lens and imagined as not being affected by this lens, is drawn as an interrupted line beyond the second lens. This indicates that the interrupted section of the ray is imaginary only.

*Step 5.* As for the optical construction, the three rules given in the previous section apply, though since an auxiliary axis is used (inclined at an angle $w$ to the main axis of the system), the following should be noted:

1. The optical axis of the system passing through the optical centres $A_1$ and $A_2$ of the two lenses, is used only to measure off distances.

Since it is assumed that all rays fall within the paraxial region, differences in length due to obliquity are negligible.

2. For the purpose of the construction, an auxiliary axis passing through the points $A_1$ and $Q_1'$ is used.

Having completed the diagram so far (*Figure 1.17*), the image $B_1'Q_1'$ and its position have been established. In order to indicate that this image is only imaginary, it too is drawn as an interrupted line. Such an image is known as a virtual image.

*Step 6.* Next the effect of the second lens must be considered. For this purpose one could imagine that the first lens did not have an effect, and that there existed only an object and the second lens (*Figure 1.18*).

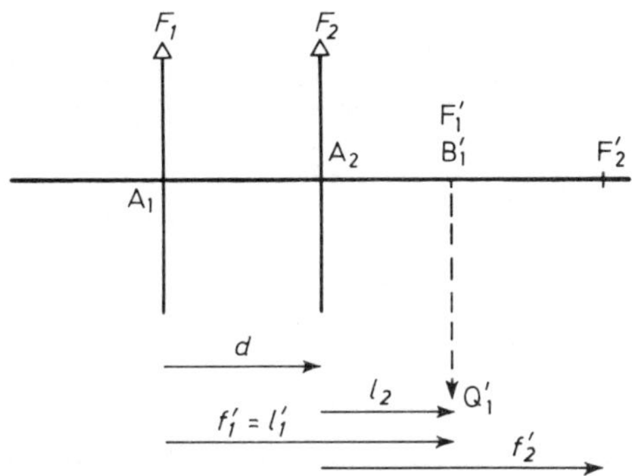


***Figure 1.18** Example 2; the first image may be considered as object for the second lens*

It will be appreciated that the image $B_1'Q_1'$ may now be considered to act as object $B_2Q_2$ (though this way of marking the 'object' is not usually used or shown in a diagram) for the second lens. Since $B_1'Q_1'$ was a virtual image, the object will be virtual too.

*Step 7.* The object distance with respect to the second lens (the new reference plane $A_2$), has to be found. Since the position of the second lens with respect to the first is known, that is, the distance between them $d = 3$ cm, the object distance $l_2$ can be calculated. From the diagram (*Figure 1.18*) it will be seen that $l_1' - d = l_2$.

NOTE

The distance $d$ is always measured from left to right, that is with the first lens as reference position.

The object distance

$$l_2 = +6.66 - (+3.00) = +3.66 \text{ cm}$$

The object vergence

$$L = \frac{100 \times 1}{+3.66} = +27.32 \text{ D}$$

*Step 8.* The conjugate foci formula applicable to the second lens, reads

$$L_2 + F_2 = L_2'$$

and the second image vergence is

$$L_2' = +27.32 + 16.66 = +43.98 \text{ D}$$

The image distance $l_2'$ is

$$l_2' = \frac{100 \times 1}{+43.98} = +2.27 \text{ cm}$$

*Step 9.* The construction of the image formed by the second lens is as follows. The auxiliary axis to be used is the one passing through the points $Q_1'$ and $A_2$. Since this is a real ray of light (*see Figure 1.20*) it is drawn as an uninterrupted line from the first lens onwards.

Having applied the third optical construction rule, one of the two remaining rules may be applied. In this case use will be made of the first principal focus of the second lens $F_2$. The position of this point is known since

$$f = \frac{-100 \times 1}{+16.66} = -6.00 \text{ cm}$$

Having indicated point $F_2$ in the diagram (*Figure 1.19*), the following reasoning is followed: a ray of light passing through $F_2$ and directed

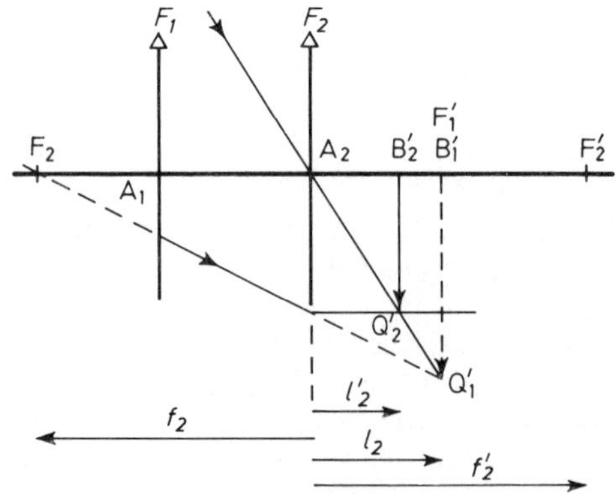


*Figure 1.19 Example 2; the image formation by the second lens*

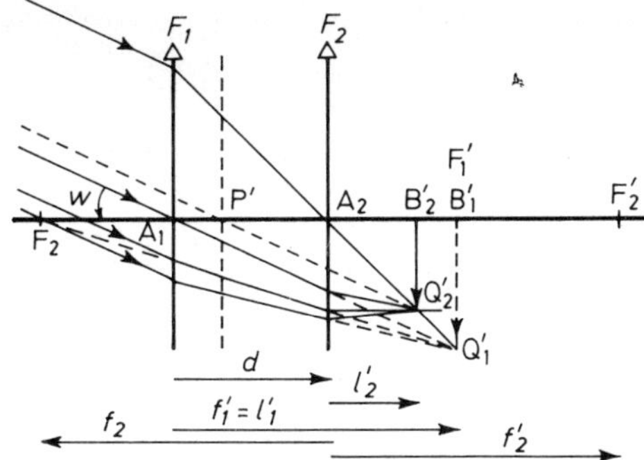


*Figure 1.20 Determination of the position of the second principal plane* P′

towards $Q_1'$, may be considered as 'a ray of light passing through the first principal focus F of a positive lens and running parallel to the optical axis after refraction'. Since the ray $F_2Q_1'$ is imaginary, it is drawn as an interrupted line. However, from the point where this ray emerges from the first lens, it is a real ray of light and is drawn uninterrupted, up to the second lens. Here, refraction takes place. The refracted real ray emerges from the second lens parallel to the optical axis $A_1A_2$ of the system. The point where this ray crosses the ray $A_2Q_1'$ marks the point $Q_2'$. This point forms the arrow head of the image. Since the image forming rays are real, the image $B_2'Q_2'$ is a real image and is thus indicated by an uninterrupted line.

Finally, these diagrams can be brought together in a composite diagram (*Figure 1.20*). It is common practice to show only the composite diagram. Having understood the principles, the reader will find with practice that such a composite diagram can be constructed without the aid of the separate intermediate stages.

In *Figure 1.20* one additional interrupted line is shown: the line $P'Q_2'$. The significance of the line is as follows. If a lens of appropriate power were placed on the optical axis $A_1A_2$ at the point P′, the distant object would also be imaged as $B_2'Q_2'$. Therefore, the second focal length of the alternative or equivalent lens system would be $P'B_2'$. This is known as the equivalent second focal length $f'$, with which is associated the equivalent focal power $F$ of the equivalent thin lens.

The point P′ is known as the second principal point of this lens system. The vertical plane through the point P′ is the second principal plane.

If the lens system would have been arranged in reversed order so that rays of light from the distant object had been incident first on the lens of +16.66 D and then on the +15.00 D lens, the first principal point P would have been determined.

The equivalent focal power measured at the point P′, may be calculated from the formula

$$F = F_1 + F_2 - dF_1F_2 \tag{1.18}$$

If the medium between the two components is not air, the distance $d$ has to be altered to its reduced distance and so equation (1.17) becomes

$$F = F_1 + F_2 - \frac{d}{n}F_1F_2 \tag{1.19}$$

NOTE

The distance $d$ must be entered in metres in equations (1.18) and (1.19). If one considers a thick lens system in air, equation (1.19) may be applied when $n$ stands for the refractive index of the medium of which the thick lens is made, and where $F_1$ and $F_2$ indicate the front and back surface powers.

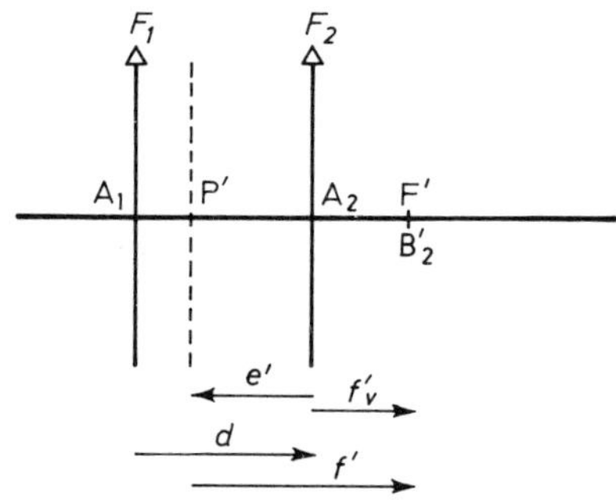


*Figure 1.21 The second principal plane* P′, *the second back vertex focal length* $f'_v$ *and the distance* $e'$

*Figure 1.21* shows the thin lens system and its second principal plane P′. The system's image plane, at $B'_2$, is also the system's second focal plane F′ since the object, in this case, was situated at infinity.

The distance from the second lens or surface to the second principal plane is denoted $e'$, that is $e' = A_2P'$. The distance from the second lens to the second equivalent focal point F′, is known as the back vertex focal length and $f'_v = A_2F'$. The back vertex power, often abbreviated to *BVP* or $F'_v$ where

$$F'_v = \frac{n'}{f'_v} \tag{1.20}$$

From *Figure 1.21* it will be seen that

$$e' = f'_v - f' \tag{1.21}$$

and that, therefore,

$$e = f_v - f \tag{1.22}$$

where $e$ is the distance from the first lens or surface to the first principal plane $A_1P$, and the distance from the first lens to the first equivalent focal point F, is $A_1P$, and denoted the front vertex focal length $f_v$. The back vertex power is also given by

$$F'_v = \frac{F}{1-(d/n)F_1} = \frac{F_1}{1-(d/n)F_1} + F_2 \tag{1.23}$$

The front vertex power *FVP* or $F_v$ is given by

$$F_v = \frac{-n}{f_v} \tag{1.24}$$

Also

$$F_v = \frac{F}{1-(d/n)F_2} = \frac{F_2}{1-(d/n)F_2} + F_1 \tag{1.25}$$

Occasionally, the distance $d$ is substituted by $t$ for the 'thickness' of a lens.

NOTE

The principal planes are not necessarily situated between $F_1$ and $F_2$, but are displaced towards the more powerful component.

Proof of the above expressions may be found in appropriate textbooks, such as Fincham and Freeman (1974).

The nodal points are two points on the axis of a lens system. A ray originating from an object point and directed at the first nodal point N, appears to leave the system through the second nodal point N$'$, towards a conjugate point of the image such that the incident and emerging rays are parallel but displaced.

Consider *Figure 1.22* which is based on *Figure 1.20*. The ray $A_1Q'_1$ originating from the distant object, is parallel to the ray $S_2Q'_2$. Before refraction by the second lens, $S_2Q'_2$ must have been directed at $Q'_1$, that is it ran along the line $S_1S_2Q'_1$. Before refraction by the first lens

this ray must have been parallel to the ray through $A_1$. If refraction would not have taken place at $S_1$, this ray would have intersected the optical axis $A_1A_2$ at N. The ray $S_2Q_2'$, if projected back, would have intersected the optical axis at N′.

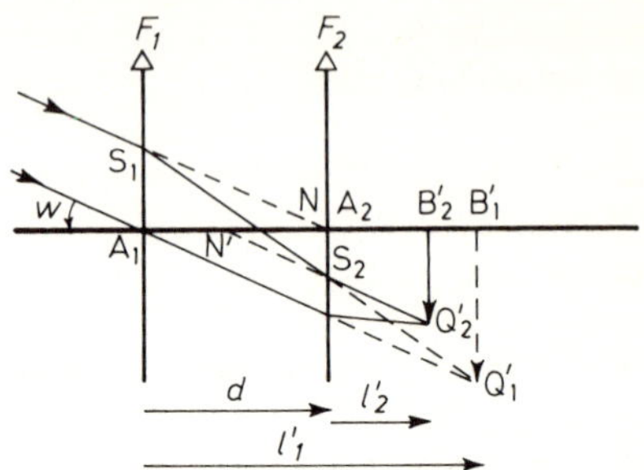


*Figure 1.22 Determination of the position of the nodal points* N *and* N′

The angles between the optical axis and the rays at $A_1$, N and N′ have the same subtense (in this example).

The position of the nodal points may be determined from the following expression (proof of which will be found in the appropriate textbooks):

$$N'F' = FP = -f \tag{1.26}$$

and

$$FN = P'F' = +f' \tag{1.27}$$

Also,

$$NN' = PP' \tag{1.28}$$

NOTE
The distance is measured from each focal point to its nodal point.

When a positive lens (system) is followed by a medium of refractive index higher than 1, the nodal points are displaced towards this medium, with respect to the principal points. The opposite holds good for a negative lens (system). When the same refractive index is present on both sides of the lens, the nodal and principal points coincide.

The importance of the nodal points is that any image size is determined by the angle subtended by the image at the second nodal point. Likewise, the object size determines the angle subtended by it at the first nodal point. The angles subtended by the object at the first and by the image at the second nodal point, are the same.

Finally, the two principal points, the two nodal points and the two focal points of a lens (system) together are known as the cardical points of the system.

## 1.3 Astigmatism

So far only spherical or stigmatic (point-image forming) pencils of light have been discussed. However, in visual optics astigmatism (not point-image forming pencils) are equally common.

Astigmatism is discussed in all textbooks on geometrical optics, and the reader is referred to these for a detailed scientific explanation of the phenomenon. However, a brief explanation follows below where it is assumed that a pencil of light from a distant point object is incident on the surface.

A spherical surface is one where the radius of curvature is constant (*Figure 1.23*). If this radius is infinitely long, $r = \infty$ and the curvature

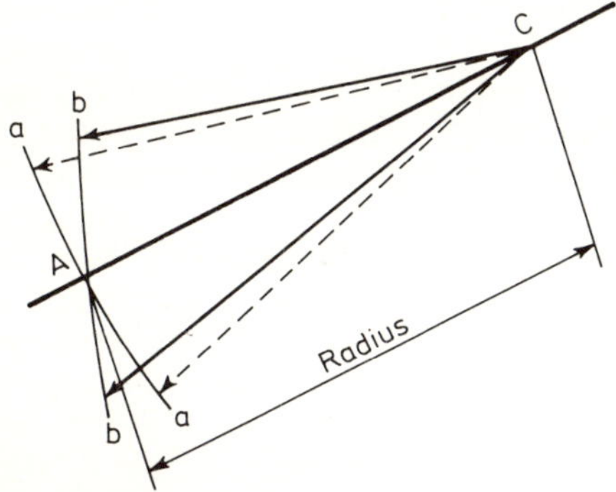


*Figure 1.23 A spherical surface*

$R = 0\ \text{m}^{-1}$. The focal power $F = 0$ D. Such a surface is referred to as being 'plano', 'flat' or afocal (not having a focus; in fact, its focus is situated at infinity).

An 'astigmatic' surface* has more than one radius of curvature. In common 'astigmatic' surfaces there are two extreme values for the

*Astigmatism refers to the image forming pencil, and not to the surface.

radius of curvature. These are to be found at right angles in the plane of the surface; the radii situated in between gradually vary from the one extreme value to the other (*Figure 1.24*). Any combination of values

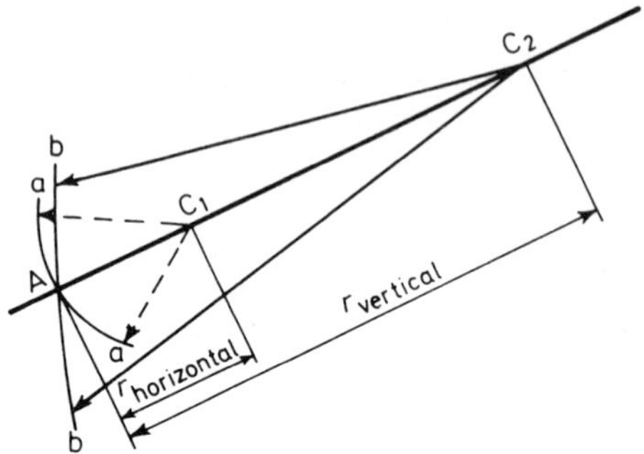


*Figure 1.24 A surface having two radii of curvature*

(both radii positive; one positive, one infinitely long; one positive, one negative; etc.) may be encountered. In the normal eye one encounters two of these cases only: both radii are positive (such as for the anterior and posterior corneal surfaces and the anterior crystalline lens surface) or both are negative (the posterior crystalline lens surface).

It will be clear that owing to the two extreme values of the radii of curvature of a surface, two extreme values for the curvature and, therefore, for the focal power and, consequently, for the focal lengths, will be found. The meridians situated between the two meridians of extreme curvature will have values varying between these two. These intermediate values are not often considered in visual optics.

In regular astigmatism the meridians showing the extreme values are at right angles to each other. These are the principal meridians. Each principal meridional power produces a focus. These foci are, owing to their power difference, at different distances from the astigmatic surface. The distance between the foci is known as the 'interval of Sturm'. Since the foci are not due to a stigmatic surface, they will not be stigmatic either, but are described as focal lines. Their formation is shown in *Figure 1.25.*

NOTE

When, for instance, the radius of curvature of one of the principal meridians lies in the vertical plane, then the power meridian will also be in the vertical plane, but the focal line due to this power will lie in the

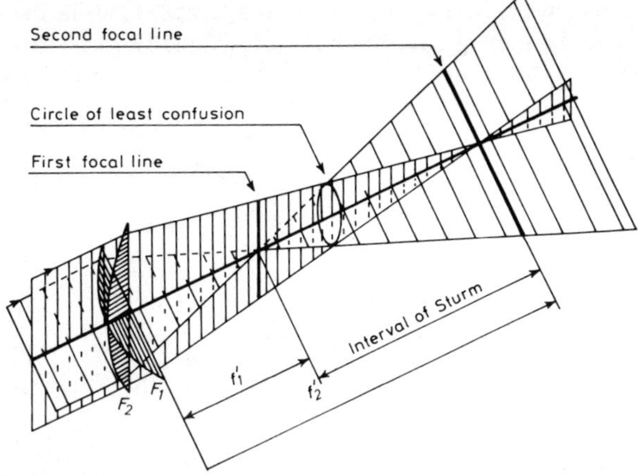


*Figure 1.25 The image formation by an astigmatic lens*

horizontal meridian. In other words, the focal lines formed by an astigmatic system are situated at right angles to the power meridians with which they are associated. Even the focal lines of so-called obliquely crossed cylinders lie at right angles to each other.

## 1.4 Optics of mirrors: catoptrics

Mirrors are plane or curved surfaces that reflect rays of light incident on them. Reflection may be regarded as a special case of refraction. Since the reflected rays are directed back into the object space with refractive index $n$, but are now travelling generally in the opposite direction (hence a change of sign), the refractive index in the image space $n' = -n$. Hence, for mirrors Snell's law (equation (1.1)) becomes

$$n \sin i \quad = \quad -n \sin i' \tag{1.29}$$

This implies that $i' = -i$ (*see Figure 1.26*), where $i$ is known as the angle of incidence and $i'$ as the angle of reflection.

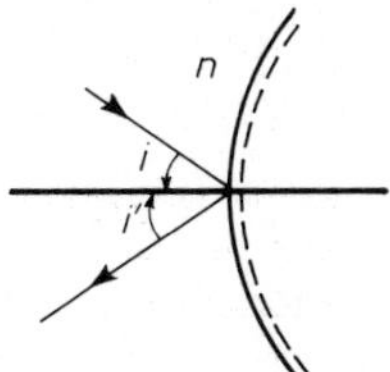


*Figure 1.26 Angle of incidence is equal to angle of reflection*

Similarly, the conjugate foci formula (equation (1.12)) may be developed as follows (employing equations (1.4), (1.10) and (1.11)):

$$\frac{n'}{l'} = \frac{n}{l} + \frac{n' - n}{r}$$

and for a case of reflection it becomes

$$\frac{-n}{l'} = \frac{n}{l} + \frac{-n - n}{r}$$

or

$$\frac{-n}{l'} = \frac{n}{l} - \frac{2n}{r} \tag{1.30}$$

Thus, the power of a mirror is given by

$$F = -\frac{2n}{r} = -\left(\frac{n}{l'} + \frac{n}{l}\right) \tag{1.31}$$

and so (*see Figure 1.27*)

$$f' = -\frac{r}{2n} \tag{1.32}$$

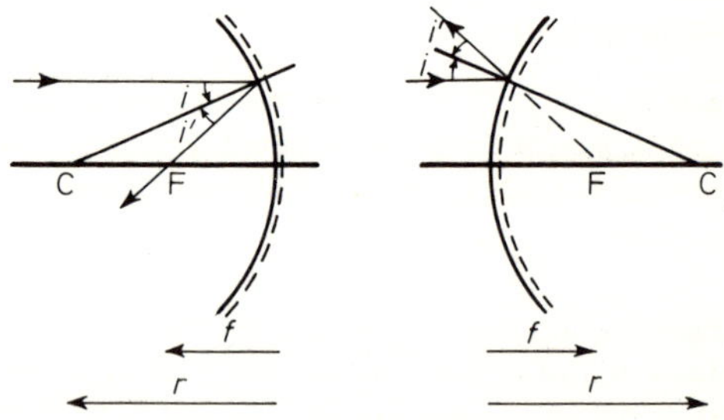


***Figure 1.27** The path of a ray incident parallel to the optical axis of a curved mirror*

Also, from expression 1.6,

$$F = \frac{-n}{f} = \frac{-n}{f'} \tag{1.33}$$

and so (*see Figure 1.27*)

$$f' = f \tag{1.34}$$

Finally, if the curvature of the mirror is known, equation (1.31) may be developed thus

$$\begin{aligned} F &= -\frac{2n}{r} \\ &= -2nR \end{aligned} \tag{1.35}$$

The magnification of an optical system is given by equation (1.15) as

$$m = \frac{L}{L'}$$

which may be developed for the case of a mirror to

$$\begin{aligned} m &= \frac{n/l}{-n/l'} \\ &= \frac{n}{l} \times \frac{-l'}{n} \\ &= \frac{-l'}{l} \end{aligned} \tag{1.36}$$

The graphical construction of an image formed by a mirror follows the same basic rules as those for an image formed by a lens. However, there are some practical differences in view of the fact that reflection, and not refraction, takes place:

1. A ray of light incident on a concave mirror parallel to its optical axis, is reflected and passes through the principal focus F, or in the case of a convex mirror, appears to originate from the focus F (*see Figure 1.27*).
2. A ray of light passing through the focus of a concave mirror follows a course parallel to the optical axis after reflection; a ray of light directed at the focus of a convex mirror follows a course parallel to the optical axis after reflection (as indicated by *Figure 1.27* when the ray paths are reversed).
3. A ray of light incident at angle $i$ at the pole, apex or vertex of a mirror, is reflected at an angle $i' = -i$ with respect to the optical axis (which coincides with the normal to the surface at the pole) (*Figure 1.26*).

A further rule which is a particular case of the last rule, applies only to reflecting surface:

4. A ray of light directed at the centre of curvature of a mirror (that is, the ray is incident along the normal to the surface), is reflected back along the path along which it was incident.

# Chapter Two
# Gullstrand's Schematic Eyes

## 2.1 Introduction

Allvar Gullstrand (1862–1930), ophthalmologist and professor of ophthalmology at Uppsala and at Stockholm, was one of the outstanding nineteenth century figures who laid the basis for our knowledge of the eye and vision in all its facets. On the basis of measurements taken by himself and others, he drew up a model eye consisting of a centred optical system having spherical surfaces. This likeness of a human eye was used to study its optical working and its performance in conjunction with spectacle lenses and optical instruments.

Following the formulation by Gauss (1840) of the laws applicable to the formation of optical images, Moser (1844) was the first to apply these to the model of an eye. However, since he chose too low a refractive index for the crystalline lens ($n$ = 1.383 9) the image plane was situated well beyond the retina, and did not represent the likeness of a human eye for the purposes mentioned in the previous paragraph.

Listing (1853) solved this problem. He introduced what is now known as 'the reduced eye'. This will be discussed in detail in section 4.2. It is similar to an imaginary eye described much earlier by Huygens (1653). He expressed the opinion that the human eye could well be represented

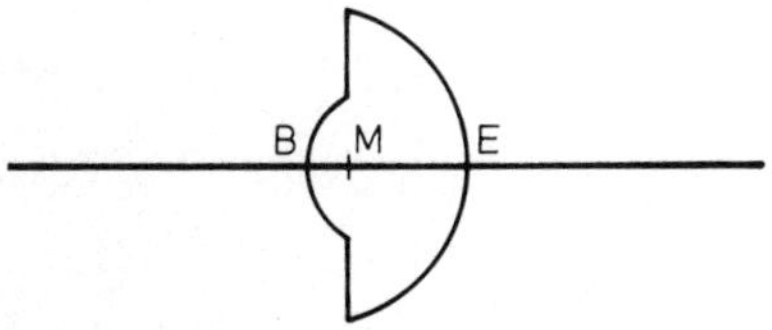

*Figure 2.1 Christiaan Huygens' simplified eye*

by two hemispheres (*Figure 2.1*). The ratio of the hemispheres could be MB ('cornea') to ME ('retina') as 1 to 3. If we assume a length of 22 mm for this eye, the radius of curvature MB would be 5.5 mm (compare with

*Table 4.1*, section 4.2). This simplified eye would be filled with aqueous humour, according to Huygens.

Helmholtz (1858), Donders (1864) as well as Gullstrand (Helmholtz, 1909–1911) adopted Listing's (1853) eye, each introducing modifications.

The importance of Gullstrand's eye lies in the fact that it gives a good idea of the average dimensions of the human eye. It must be stressed, however, that these are only average values arrived at on a mathematical basis or by way of reason. They are not measurements taken from one existing eye. In short, this kind of eye is wholly fictitious and may serve only as a means of communicating the optical principles involved.

Below follows an outline of the development of Gullstrand's eyes. The full account may be found in Helmholtz' *Physiologische Optik* (1909–1911), or in Southall's (1924) English translation. Accounts of more recently made measurements and observations have been added. See also *Figures 4.1* and *4.2.*

## 2.2 The cornea

### 2.2.1 THE ANTERIOR RADIUS

The human cornea is not a spherical surface having one radius of curvature. Its surface is rather complex, even if the likely presence of astigmatism is omitted for the sake of convenience.

The central portion which is known as the cap is a zone of about 5 mm diameter. The radius of curvature in this zone varies but little. The cap is surrounded by an annulus of about 3 mm width where the radius of curvature is longer than that of the cap by about 0.5 mm. This annulus is surrounded by a second contained by the circumference of the first and the limbus. In the second annulus the radius of curvature is shorter than in the first. Besides, different curvatures are found along different meridians. Gullstrand was aware of the peripheral flattening of the cornea and the variations along different meridians. However, he thought that these irregularities were due to unavoidable errors of observation. Modern methods have shown that this is not so (Le Grand, 1962). Other factors that should be mentioned are:

1. The average cap is decentred. Helmholtz and his contemporaries stated that this was temporally and inferiorly, but Bier (1956) wrote that it was nasally and superiorly. In any case, these decentrations are less than 1 mm.

2. The corneal surface is subject to diurnal variations. This affects mainly the area along the cap's circumference (Le Grand, 1962).
3. This same area is subject to variations induced by the ambient temperature. According to Le Grand (1962), a rise in temperature causes flattening.
4. Menstruation and ovulation affect the corneal curvature, too (Le Grand, 1962).

Gullstrand decided that Donders' (1864) data obtained with a Helmholtz 'ophthalmometer' (keratometer) were to be used since these were in good agreement with those of other researchers such as Steiger (1895), with respect to the '*ophthalmometric average value* of the radius of curvature of the optical zone of the cornea'. Donders calculated an average radius of curvature of the cornea of 7.8 mm, with physiological limits of 7 and 8.5 mm. He found slightly steeper radii in women than in men. This was confirmed by Steiger.

Bier (1957) stated that the average radius of the cap was to be found between 7.70 and 8.10 mm. Le Grand (1964) considered 7.8 mm to be a reasonable average, and Westerhout's (1970) graphs tend to bear out these figures.

Gullstrand adopted a *schematic value* of 7.7 mm (*see* section 2.2.4).

It should be pointed out that the radius of curvature measured with a keratometer is not that of the anterior surface of the cornea, but of the anterior surface of the tears liquid layer which normally covers the cornea. Gullstrand ignored this layer saying it could be considered as an infinitely thin film bounded by concentric surfaces. Therefore, it amounts to an afocal component which does not affect the optical ray path.

### 2.2.2 THE REFRACTIVE INDEX

Following the same reasoning, no account needs to be taken of the refractive index of the tears liquid layer. Its refractive index is usually taken to be 1.336 (Bennett, 1956; Stone and Phillips, 1972).

The introduction of Abbe's refractometer made it possible to measure the refractive index of the cornea with accuracy. Aubert and Matthiessen (Aubert, 1876) found 1.377 for a 50 years' old cornea of a male, and 1.372 1 for the cornea of a two days' old child. Matthiessen (1891) decided upon 1.376 3 whereupon Gullstrand judged 1.376 to be the best *schematic value.*

Bier (1957) gave this latter figure as did Stone (Stone and Phillips, 1972). Le Grand (1964) showed the value of 1.377 1 measured for a wavelength of approximately 550 nm.

### 2.2.3 THE THICKNESS

Blix, a fellow-countryman of Gullstrand, developed an elegant method of measuring the corneal thickness (Blix, 1880). The eye piece of a microscope is replaced by an illuminated object (target) which is imaged by the high power objective of the microscope. A second microscope is used by the observer to take measurements. This microscope is inclined to the first in such a way that the two optical axes cross at the point where the image is formed by the first microscope. The observer's microscope should be adjusted to observe this image when reflected from a surface. The two microscopes are moved in unison, along the normal to the corneal surface whereby the image is reflected by the anterior surface. The difference in position between this image and that obtained by reflection from the posterior surface of the cornea, is a measure of the apparent corneal thickness. The real corneal thickness may be calculated by making allowance for the refractive index of the cornea.

Blix found normal limits of 0.506 and 0.576 mm in the measurements of ten eyes. Gullstrand himself measured the corneal thickness of the eyes of two subjects by means of a slit lamp. He found values of 0.46 and 0.51 mm. He, therefore, decided to adopt Blix's proposal that the *schematic value* of the thickness of the cornea be taken as 0.5 mm.

Norn's (1974) collection of corneal thickness measurements agrees with the table given by Duke-Elder (1970) and showed that the more recent measurements were between 0.52 and 0.53 mm.

### 2.2.4 THE POSTERIOR RADIUS

Because of the 'thinness' of the cornea and of its refractive index, any light reflected from the posterior surface of the cornea is almost completely obscured by light reflected from the anterior surface. In addition, the amount of light reflected by the corneal–aqueous interface is very little, owing to the very small difference in refractive index ($n_{\text{cornea}} = 1.376$ and $n_{\text{aqueous}} = 1.336$).

The following calculations will give an idea of the relative luminances of the two corneal reflections.

The reflection factor may be calculated from Fresnel's formula which reads (for small angles of incidence)

$$\rho = \left(\frac{n' - n}{n' + n}\right)^2 \tag{2.1}$$

The reflection factor at the corneal–aqueous (as well as at the tears liquid–corneal) interface is

$$\rho = \left(\frac{1.376 - 1.336}{1.376 + 1.336}\right)^2 = 0.014\,74^2 \approx 0.000\,2$$

The reflection factor at the air–corneal interface is

$$\rho = \left(\frac{1.336 - 1}{1.336 + 1}\right)^2 = 0.143\,8^2 \approx 0.02$$

(For the sake of accuracy the refractive index of the tears liquid has been used instead of that of the cornea; however, the difference is negligible.)

It will be seen from these calculations that the difference in luminance of the corneal reflections is about one hundred fold. This makes it very difficult to detect the reflection from the posterior surface.

One way of taking such measurements is to work in the dark and at an angle to the normal of the cornea. In this way the corneal thickness is better seen [illegible] two reflections become separated. Allowance will have to be made for the angle of observation. Graves (1968) suggested immersion of the cornea in water. The result would be that the two reflections would become more equal in brightness.

The reflected images described above are known as Purkinje–Sanson images numbers I (from the anterior corneal surface) and II (from the posterior surface), and these will be discussed in chapter 19. They were used by Gullstrand to determine the posterior radius of curvature of the cornea (by comparison with those of the anterior surface). Gullstrand concluded that the *ophthalmometric mean value* of the radius of the posterior surface (of the optical zone or cap) could hardly be greater than 6.7 mm.

Gullstrand continued by saying that in view of inaccuracies in the measuring technique and uncertainties concerning the exact corneal areas measured, he assumed *schematic values* of the radii of curvature at the vertices of the optical zone of the cornea of 7.7 and 6.8 mm.

Apart from Tscherning's (1898) values of between 6.2 and 6.6 mm, Thomson (cited by Thomas, 1955) recorded 6.8 and Cogan (1951) 6.6 mm as radius of curvature of the posterior surface of the cornea.

### 2.2.5 THE REFRACTIVE INDEX OF THE AQUEOUS AND VITREOUS HUMOURS

Although different investigators obtained slightly varying values, Gullstrand did not judge it necessary to adopt a value different from that

obtained by Helmholtz with the refractometer. The latter gave 1.336 5 as refractive index of aqueous humour and 1.338 2 of vitreous humour. Gullstrand assumed for both humours the *schematic value* of 1.336.

Graves (1968) gave for a wavelength of 550 nm at 37 °C (human body temperature) the refractive index of the aqueous as 1.337 4 and of the vitreous as 1.336 0.

### 2.2.6 THE CORNEAL SYSTEM

From the preceding paragraphs the following table of constants for the cornea may be compiled:

| | *Ophthalmometric mean values* | *No. 1 Exact Schematic Eye* |
|---|---|---|
| Anterior radius, $r_1$ | +7.8 mm | +7.7 mm |
| Posterior radius, $r_2$ | +6.7 mm | +6.8 mm |
| Thickness, $t$ | +0.5 mm | +0.5 mm |
| Refractive index cornea, $n_c$ | 1.376 | 1.376 |
| Refractive index aqueous, $n_a$ | 1.336 | 1.336 |

On the basis of these data the following ophthalmometric values may be calculated:

Anterior surface power (equation (1.4))

$$F_1 = \frac{n_c - 1}{r_1} = \frac{(1.376 - 1) \times 1000}{+7.8} = \frac{376}{+7.8} = +48.21 \text{ D}$$

Posterior surface power

$$F_2 = \frac{n_c - n_a}{-r_2} = \frac{(1.376 - 1) \times 1000}{-(+6.7)} = \frac{40}{-6.7} = -5.97 \text{ D}$$

Front vertex power (equation (1.24))

$$F_v = \frac{F_2}{1 - \frac{t}{n_c}F_2} + F_1 = \frac{-5.97}{1 - \frac{0.005}{1.376}(-5.97)} + 48.21 = +42.36 \text{ D}$$

Front vertex focal length (equation (1.8))

$$f_v = -\frac{1}{F_v} = -23.61 \text{ mm}$$

Back vertex power (equation (1.22))

$$F'_v = \frac{F_1}{1 - \frac{t}{n_c}F_1} + F_2 = \frac{+48.21}{1 - \frac{0.005}{1.376}(+48.21)} + (-5.97) = +52.48$$

Back vertex focal length (equation (1.9))

$$f'_v = \frac{1}{F'_v} = +19.05 \text{ mm}$$

Equivalent focal power of the corneal system (equation (1.18))

$$F = F_1 + F_2 - \frac{t}{n_c} F_1 F_2 = +48.21 - 5.97 - \frac{0.005}{1.376}(+48.21)(-5.97)$$

$$= +43.28 \text{ D}$$

Equivalent focal lengths of the corneal system (equation (1.6))

$$f' = -f = \frac{1}{F} = +23.10 \text{ mm}$$

First principal point distance (equation (1.21))

$$e = f_v - f = -23.61 - (-23.10) = -0.51 \text{ mm}$$

Second principal point distance (equation (1.20))

$$e' = f'_v - f' = +18.93 - (+23.10) = -4.17 \text{ mm}$$

Similarly, for the No.1 Exact Schematic Eye:

$$F_1 = +48.83 \text{ D}$$
$$F_2 = -\ 5.88 \text{ D}$$
$$F_v = +43.07 \text{ D}$$
$$F'_v = +53.67 \text{ D}$$
$$F = +43.99 \text{ D}$$

$$f_v = -23.22 \text{ mm}$$
$$f'_v = +18.63 \text{ mm}$$
$$f' = -f = +22.73 \text{ mm}$$
$$e = -0.49 \text{ mm}$$
$$e' = -4.10 \text{ mm}$$

It will be seen from these figures that these variations make but little difference to the overall values. Graves (1968) expressed this view in saying that all reasonable choices lead to neighbouring results.

## 2.3 Depth of the anterior chamber

The depth of the anterior chamber should be defined as the distance between the apex of the posterior corneal surface and the apex of the anterior surface of the crystalline lens. Some measurements were made using the edge of the iris rather than the anterior surface of the lens.

According to Gullstrand the best method to measure the depth of the anterior chamber is Blix's method (*see* section 2.2.3). Unfortunately, he measured only one emmetropic eye at the corneal vertex (mean value 3.515 mm) while another four emmetropic eyes were measured along the visual axis. However, many other investigators measured this depth using other methods. Their results are all so much alike that Gullstrand thought it justified to assume a *schematic value* of 3.6 mm (lens unaccommodated).

Sorsby *et al.* (1957) found that the extreme values for the depth are 2.7 and 4.5 mm, and showed that the peak of the distribution curve which fitted the data best lay between 3.4 and 3.6 mm.

## 2.4 The crystalline lens

### 2.4.1 THE RADII OF CURVATURE

Gullstrand sufficed to say that Helmholtz' assumed *schematic value* of 10 mm for the anterior radius and 6 mm for the posterior radius, fell within the limits of the errors involved in the method of measurement. The method used is known as ophthalmophakometry. It employs a modified keratometer to measure the radii of curvature of the crystalline lens (section 19.3).

Emsley (1955) stated that the value for the anterior radius is nearer 11 and 10 mm. Duke-Elder (1961) remarked that anatomical measurements are probably very inexact and that more reliance should be placed on optical measurements in the living eye. For the anterior radius Duke-Elder gave values varying from 8.4 to 13.8 mm with an average of 10 mm, and for the posterior radius values varying from 4.6 to 7.5 mm with an average of 6 mm.

### 2.4.2 THE REFRACTIVE INDEX

The measurement of the refractive index of the crystalline lens poses great problems:

1. Direct measurements can only be made after extraction of the lens. However, at that time the lens is no longer in its natural environment and this is likely to introduce discrepancies.

2. The refractive index of the crystalline lens is not uniform, but varies. Gullstrand wrote that physiological measurements indicated that until puberty the variation of the refractive index is continuous throughout the lens. Around the age of 30 discontinuity in the variation of the refractive index becomes manifest as the iso-indical surfaces, each of which being the locus of points where the refractive index has the same value. These may be observed with a slitlamp.

Freytag (1907) carried out measurements with a refractometer. Gullstrand used the following of these measurements as schematic values:

| | | | |
|---|---|---|---|
| Anterior vertex | 1.387 | Posterior vertex | 1.385 |
| Equator | 1.375 | Centre of the lens | 1.406 |

Gullstrand then assumed a value of 1.386 as *schematic value* of the index of refraction of the two vertices.

Gullstrand described the efforts made to obtain very accurate details about the lens' indices, introducing such notions as the total refractive index and the equivalent core lens. The latter is defined as a lens having the same refractive index as that at the centre of the crystalline lens (1.406) and suspended in a medium with the same refractive index as that at the vertices of the crystalline lens (1.386), its refractive power and the positions of its principal points being identical with those of the real core lens. The radii and thickness of the equivalent core lens are found to be:

$$r_1 = +7.910\,8 \text{ mm}$$
$$r_2 = -5.760\,5 \text{ mm}$$
$$t = +2.418\,7 \text{ mm}$$

However, as Le Grand (1964) points out, by the time Gullstrand had reached this point in his essay so many assumptions had been made that the whole concept had become rather arbitrary and fictitious. The reader is given the impression of rigorous application of mathematics, but this does not stand up to detailed scrutiny, as was shown by Huggert (1946), cited by Le Grand (1964).

Graves (1968) gave the following values for a wavelength of 550 nm and 37 °C:

| | | | |
|---|---|---|---|
| Lens capsule | 1.360 | Pole | 1.387 |
| Equator | 1.375 | Centre of the lens | 1.41 |

and added that one needs a homogeneous lens of refractive index 1.42 in order to replace the crystalline lens and retain the same properties.

### 2.4.3 THE REFRACTIVE POWER

Gullstrand took, on the one hand, the power of the correcting spectacle lenses for eyes of which the crystalline lens had been removed by operation, as basis for his calculations of the refractive power of the crystalline lens, and on the other hand, an approximate law of the increase of the refractive index after Matthiessen (1877). In this way he arrived at a refractive power of +19.110 7 D.

Sorsby *et al.* (1957) gave a mean value of +19.7 D with a standard deviation of 1.62 D.

Graves (1968) gave the power of the crystalline lens as +21.778 7 D, on the basis of an anterior radius of +10.2 mm, a posterior radius of −6.0 mm and a refractive index of 1.42.

## 2.5 The optical system of the eye

*Table 2.1* – the Schematic Eye – shows the data of such an eye based on the equivalent core lens (*see* section 2.4.2). The data are given to three decimal places and may, therefore, show slight differences with values given above.

In *Table 2.1* the 'position of the retinal fovea' appears for the first time without introduction. We also find the information that there is 'hypermetropia along the axis' (*see* section 3.3.2.1) of 1 D. These points are then discussed. The axial length of 24 mm is just within the limits that Treutler (1902) thought to be reasonable. Allowing for a 1 mm thickness of sclera and choroid, Mauthner's (1876) measurements on dead eyes gave values around 25 mm which would appear to be in good agreement.

Sorsby *et al.* (1957) gave a mean value and standard deviation for the axial length of the eye of 24.4 mm ± 0.85. Le Grand (1964) calculated the axial length of his theoretical eye (*see* section 4.3) to be 24.20 mm. Graves (1968) said that in view of recent measurements in living eyes he adopted the value of 24 mm for the distance between the apex of the cornea and the macula.

Having discussed the effect aberrations have on the dioptric power of the eye, Gullstrand concluded that it is absolutely necessary to take these aberrations into account in the Exact Schematic Eye. In view of the difficulties which this posed he suggested that it might be desirable

TABLE 2.1

Gullstrand's schematic eyes

| | *No.1 Exact* | | *No.2 Simplified* | |
|---|---|---|---|---|
| | *Unaccommodated* | *Max. accommodated* | *Unaccommodated* | *Max. accommodated* |
| Refractive index | | | | |
| cornea | 1.376 | 1.376 | | |
| aqueous and vitreous | 1.336 | 1.336 | 1.336* | 1.336* |
| lens | 1.386 | 1.386 | †1.413* | 1.424* |
| equivalent core lens | 1.406 | 1.406 | | |
| Position (mm) | | | | |
| anterior corneal surface | 0 | 0 | 0 | 0 |
| posterior corneal surface | +0.5 | +0.5 | | |
| anterior lens surface | +3.6 | +3.2 | | |
| equiv. core lens' ant. surface | +4.146 | +3.872 5 | | |
| equiv. core lens' post. surface | +6.565 | +6.527 5 | | |
| posterior lens surface | +7.2 | +7.2 | | |
| optical centre of lens | | | +5.85* | +5.2* |
| Radius of curvature (mm) | | | | |
| anterior corneal surface | +7.7 | +7.7 | | |
| posterior corneal surface | +6.8 | +6.8 | | |
| equivalent corneal surface | | | +7.8* | +7.8* |
| anterior lens surface | +10.0 | +5.33 | +10.0* | +5.33* |
| equiv. core lens' ant. surface | +7.911 | +2.655 | | |
| equiv. core lens' post. surface | −5.76 | −2.655 | | |
| posterior lens surface | −6.0 | −5.33 | −6.0* | −5.33* |
| Refracting power (D) | | | | |
| anterior corneal surface | +48.83 | +48.83 | | |
| posterior corneal surface | −5.88 | −5.88 | | |
| equivalent corneal surface | | | +43.08 | +43.08 |
| anterior lens surface | +5.0 | +9.375 | +7.7 | +16.5 |
| core lens | +5.985 | +14.96 | | |
| posterior lens surface | +8.33 | +9.375 | +12.833 | +16.5 |

| | No.1 Exact | | No.2 Simplified | |
|---|---|---|---|---|
| | *Unaccommodated* | *Max. accommodated* | *Unaccommodated* | *Max. accommodated* |
| Corneal system | | | | |
| refracting power (D) | +43.05 | +43.05 | +43.08 | +43.08 |
| position first principal point (mm) | −0.049 6 | −0.049 6 | 0 | 0 |
| position second principal point | −0.050 6 | −0.050 6 | 0 | 0 |
| first focal length (mm) | −23.227 | −23.227 | −23.214 | −23.214 |
| second focal length | +31.031 | +31.031 | +31.014 | +31.014 |
| Lens system | | | | |
| refracting power (D) | +19.11 | +33.06 | +20.53 | +33.0 |
| position first principal point (mm) | +5.678 | +5.145 | +5.85 | +5.2 |
| position second principal point | +5.808 | +5.255 | +5.85 | +5.2 |
| focal length (mm) | +69.908 | +40.416 | +65.065 | +40.485 |
| Complete optical system of eye | | | | |
| refracting power (D) | +58.64 | +70.57 | +59.74 | +70.54 |
| position first principal point (mm) | +1.348 | +1.772 | +1.505 | +1.821 |
| position second principal point | +1.602 | +2.086 | +1.631 | +2.025 |
| position first focal point | −15.707 | −12.397 | −15.235 | −12.355 |
| position second focal point | +24.387 | +21.016 | +23.996 | +20.963 |
| first focal length | −17.055 | −14.169 | −16.740 | −14.176 |
| second focal length | +22.785 | +18.930 | +22.365 | +18.938 |
| position fovea centralis (mm) | +24.0 | +24.0 | +24.0 | +24.0 |
| axial refraction (D) | +1.0 | +9.6 | 0 | +9.7 |
| position near point (mm) | | −102.3 | 0 | −100.8 |

*Assumed data
†Different authors give different values; some examples are:

| | | | |
|---|---|---|---|
| Emsley (1955) | 1.416 | Schober (1970) | 1.413 |
| Graves (1968) | 1.42 | Southall (1924) | 1.413 |
| Le Grand (1964) | 1.4085 | | |

to calculate a Simplified Schematic Eye where the divergent effect of the posterior corneal surface is ignored, the crystalline lens is homogeneous, and the optical system is free from aberrations. Having gone that far one might as well assume the position of the optical centre of the lens. The next step was then to ignore the principal points, and to replace them by the equivalent surface of the cornea. The radius of curvature of this surface agrees very closely with the schematic radius of the optic zone as measured with a keratometer. Consequently, this value was assumed. Although Gullstrand did not say so, the fact that he made all these assumptions is a further justification for the reduction of the number of decimal places of the 'assumed data' (*Table 2.1*) so that one is no longer given the impression that all measurements are of such a high degree of accuracy.

On the basis of the simpler data the values of the other optical entities of the eye could be calculated.

Gullstrand added the following remarks:

1. A reduced eye (*see* section 4.2) corresponding to the Exact Schematic Eye and having a refractive index of 4/3, would have a radius of curvature of 5.7 mm.
2. In the Simplified Eye the total index of the lens is greater than its actual behaviour indicates.
3. Therefore, the Exact Eye should be used for ray tracing purposes.
4. In the Schematic Eye the decentration of the refracting surfaces as well as the obliquity of the line of sight (the visual axis; section 19.4) have been ignored because of their variability and non-sphericity. In fact, this optical system has no axis of symmetry, yet investigations have shown that the image formation in the eye is not affected.
5. The retina is adapted for distinct image reception over a very small area only (the fovea centralis). Though the foveal image belongs, really, to the peripheral part of the image since the visual and optical axes (section 19.4) do not coincide, the practical effects are so good that, as a first approximation, the optical system may be considered as a centred system.

# Chapter Three
# Emmetropia and Ametropia

## 3.1 Introduction

The refractive state of an eye, that is, whether it does or does not require an optical correction (in principle), is always defined with respect to infinity.

## 3.2 Emmetropia

When an eye is said to be emmetropic (from the Greek: en = in metron = measure, ops = to see; loosely translated as 'in correct seeing adjustment'), it means that a distant object (in principle, an object that is infinitely far away from that eye) is imaged on its retina. Expressed in optical terminology, the distant object is conjugate with the eye's retina by means of its optical system.

It should be noted that this does not necessarily mean that the person whose eye is emmetropic will be aware of the object when it is imaged on his eye's retina. If the visuum (a collective phrase meaning all nerve pathways concerned with vision) does not function normally, the person may not 'see' the object, or may not see it clearly. In the latter case one denotes the condition as amblyopia (Greek amblus = dull, blunt). This condition cannot be improved by means of an optical aid to reach 'normal' vision. If the eye's nervous system does not function at all, the eye will be blind yet emmetropic. Hence emmetropia is a term applied to the purely optical side of the visual process. Emmetropia is, therefore, sometimes referred to as 'the condition of ideal focus'. Furthermore, for the image to be perceived clearly it should be formed on the macular area of the retina. This is the area of the retina where vision is most acute.

## 3.3 Ametropia

This is the opposite of emmetropia (a = not, without), that is, a distant object is not imaged by the eye's optical system on the retina, but the image is formed in front of the retina, or would be formed behind the retina if the latter were removed.

Since the refractive state is defined with respect to infinity, it is always assumed that the eye is in its unaccommodated state (*see* chapter 12).

### 3.3.1 CLASSIFICATION

Ametropia may be classified as follows:

*Spherical* (stigmatic) ametropia
- axial
- refractive; subdivisions:
  - index
  - curvature
  - position of an element
  - absence of an element

*Astigmatic* ametropia or *astigmatism*
- index
- curvature
- position of an element

### 3.3.2 SPHERICAL OR STIGMATIC AMETROPIA

If an eye were emmetropic it would form a point image of a distant point object (stigma = point) on its retina. However, an ametropic eye forms a point image of a distant point object, but not on its retina. Instead a blur circle (*see* section 8.2) is formed on its retina.

It is customary to subdivide ametropia into a number of conditions for the sake of convenience. Whether or not a particular eye is subject to such a condition is usually not known. This is because an investigation into the cause of the ametropia is very complicated and time consuming. Besides, with respect to its optical correction, the cause of the ametropia is of importance in only a limited number of cases. Moreover, ametropia may be due to a number of causes which, together, produce the condition. However, for the purpose of visual optics calculations it is convenient to use this classification.

#### *3.3.2.1 Axial ametropia*

The second principal focus of the eye and the retina do not coincide: the eye's axial length is either too short or too long with respect to its second focal length.

### *3.3.2.2 Refractive ametropia*

The second principal focus of the eye does not coincide with the retina: in this case the second focal length of the eye is either shorter or longer owing to a departure from the normal power of the eye's optical system with respect to the normal axial length.

In both axial and refractive ametropia the reduced eye is usually considered to have a power of +60.00 D and the associated axial length is +22.22 mm.

NOTE
It should be remembered that this method of defining axial and refractive ametropia, that is, with respect to +60.00 D power and +22.22 mm axial length, is only a technical tool useful in visual optics. In human eyes neither value need apply.

*Index ametropia* denotes an anomaly in the refractive index of either cornea, aqueous humour, crystalline lens or vitreous body, or a combination of these. The most likely component to show (a) variation(s) is the crystalline lens.

*Curvature ametropia* refers to an anomaly in the curvature of the refracting surfaces of the cornea and crystalline lens, that is, steeper or flatter than the normal curvature. This may result in a departure from the normal power of the eye.

*Position of an element*, such as a displacement of the crystalline lens, can cause ametropia. The lens may be displaced towards or away from the cornea. If the lens is partially displaced upwards, or more commonly, downwards, it is known as subluxation of the lens.

*Absence* of the crystalline lens causes a condition known as aphakia (*see* chapter 14).

## 3.3.3 ASTIGMATIC AMETROPIA

In this case the optical system of the eye does not form a point image of a distant point object (a = not; stigma = point), but two mutually perpendicular line foci separated by a certain distance (*see* section 1.3).

*Index astigmatism* may occur when the crystalline lens shows variations in its refractive index in different meridians.

*Curvature astigmatism* is present when any of the refracting surfaces is not spherical, but shows different radii of curvature along different meridians.

The *position of an element*, that is, when the crystalline lens is inclined to the optical axis of the system, may cause astigmatism.

Finally, the posterior pole of the eye may be inclined with respect to the axis of the system. This condition does not give rise to astigmatism which can be corrected by means of an astigmatic lens since this is not a departure from the normal eye's optical system, but of the position of the retinal receptor layer. It is sometimes referred to as retinal obliquity.

## 3.4 Clinical aspects of ametropia

### 3.4.1 CLASSIFICATION

For clinical purposes ametropia is classified as follows:

Myopia
Hyperopia (hypermetropia)
Astigmatism

### 3.4.2 MYOPIA

The meaning of this Greek word is 'I close the eye'. This refers to the habit of myopes of closing their eyes partially, thus producing a pinhole effect (*see* section 8.2) which improves the quality of the retinal image.

Myopia is also known as shortsightedness. This term is due to the fact that the myope can see objects clearly that are a relatively short distance away from his eyes. An object at infinity, however, will always be

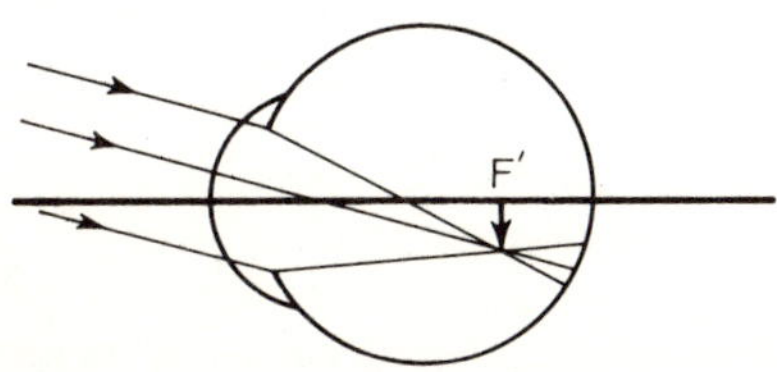


*Figure 3.1 The image formation of a distant object in a myopic eye*

imaged out-of-focus unless the myope has been corrected by an optical aid. A distant object will be imaged in front of the myopic eye's retina (*Figure 3.1*). This indicates that the axial length of the myopic eye is too long relative to its second focal length.

The amounts or degrees of myopia are usually clinically classified as follows:

| | | | |
|---|---|---|---|
| Low | 0.25 to 3.00 D | High | 6.25 to 10.00 D |
| Medium | 3.25 to 6.00 D | Very high | over 10.00 D |

'Very high' myopia is often considered to be a pathological condition.

This classification, and those for hyperopia and astigmatism, should be looked upon only as a rough indication of the degree of ametropia to be used for descriptive purposes.

### 3.4.3 HYPEROPIA (HYPERMETROPIA)

The original term, hypermetropia (hyper = above, in excess), appears to be less frequently used today than its abbreviated form hyperopia which was coined by Helmholtz.

Hyperopia is also known as farsightedness or longsightedness. This term is due to the fact that hyperopes can usually see objects that are relatively far away from their eyes, (quite) clearly.

In a hyperopic eye a distant object would be focussed behind the retina if the latter were not present (*Figure 3.2*). However, since the

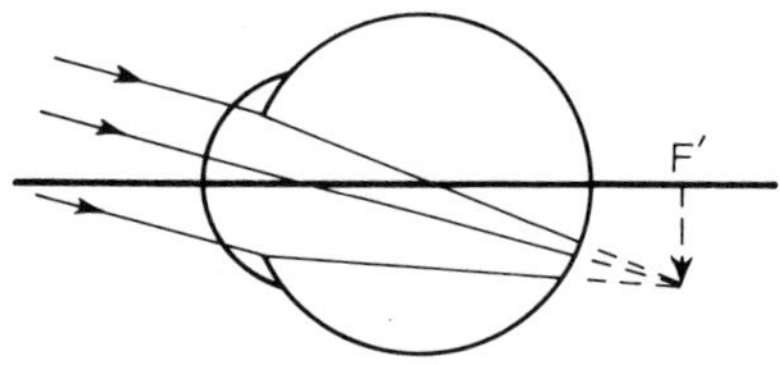


*Figure 3.2 The image formation of a distant object in a hyperopic eye*

retina is present in a normal eye, the retinal image will be out-of-focus or blurred. This indicates that the axial length of the hyperopic eye is too short relative to its second focal length. However, there is a process known as accommodation which may assist the uncorrected hyperope in focussing an object (distant or near) on the retina. This will be discussed in chapter 12.

Hyperopia is usually classified for clinical purposes into

| | |
|---|---|
| Low | 0.25 to 3.00 D |
| Medium | 3.25 to 5.00 D |
| High | over 5.25 D |

### 3.4.4 ASTIGMATISM

This is classified as follows (*Figure 3.3*):

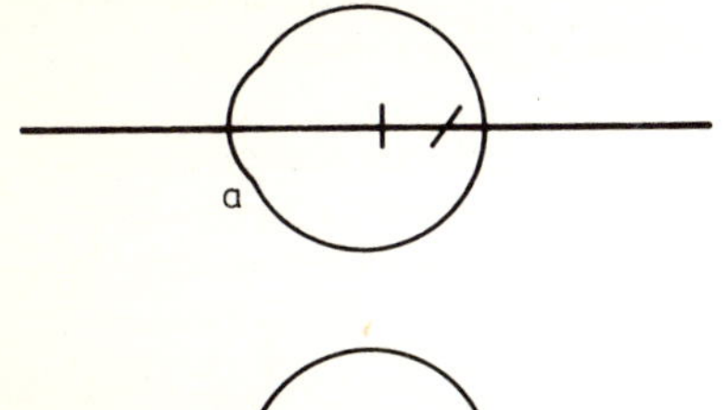


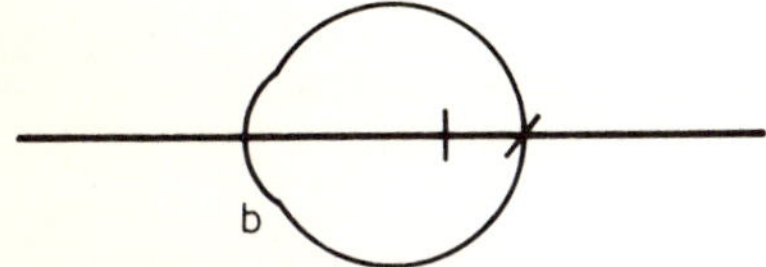


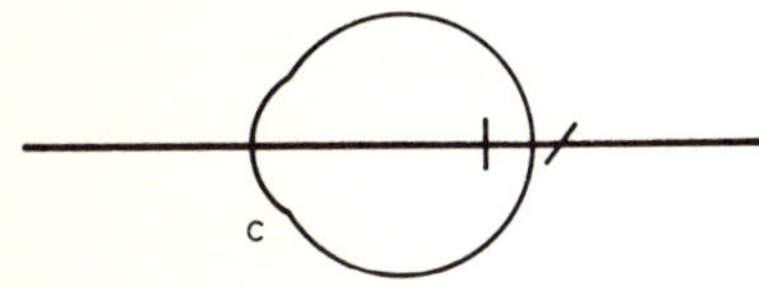


*Figure 3.3 Positions of the focal lines in: (a) compound myopic astigmatism, (b) simple myopic astigmatism; (c) mixed astigmatism, (d) simple hyperopic astigmatism, and (e) compound hyperopic astigmatism*

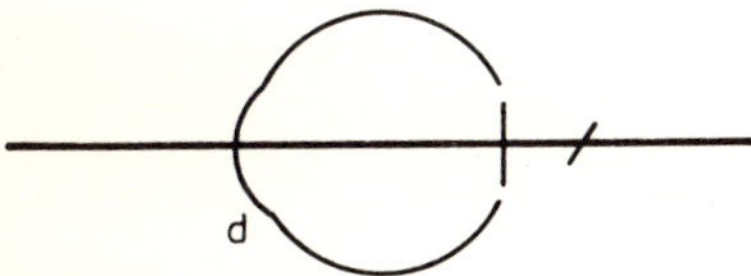


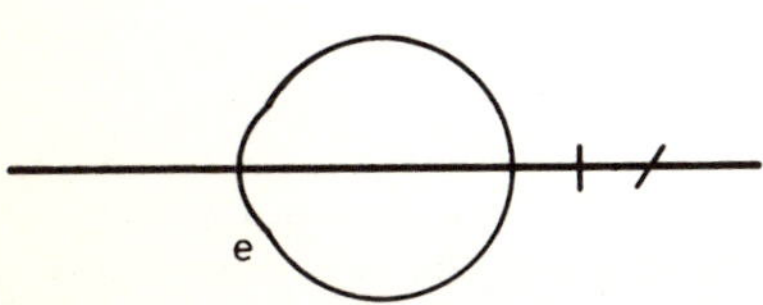


(a) Compound myopic astigmatism – both focal lines are formed in front of the retina

(b) Simple myopic astigmatism – the first focal line is formed in front of, the second on the retina

(c) Mixed astigmatism – the first focal line is formed in front of, the second behind the retina

| | |
|---|---|
| (d) Simple hyperopic astigmatism | – the first focal line is formed on the retina, the second behind it |
| (e) Compound hyperopic astigmatism | – both focal lines would be formed behind the retina |

Another classification of astigmatism is also used: this specifies the direction of the astigmatism. If the more powerful (positive) meridian of the ocular astigmatism is (approximately) vertical, it is known as 'with the rule', 'direct' or as 'astigmatismus rectus' since this is most often encountered amongst astigmatic eyes.

If the more powerful meridian is situated (approximately) horizontally it is known as 'against the rule', 'indirect' or as 'astigmatismus inversus'. However, if the more powerful meridian is neither (approximately) horizontal nor vertical, but is situated somewhere in between, it is known as 'oblique astigmatism'.

As for the degrees of astigmatism these are usually divided for clinical and descriptive purposes into

| | |
|---|---|
| Low | 0.25 to 1.00 D |
| Medium | 1.25 to 3.00 D |
| High | over 3.25 D |

where the dioptric power indicates the difference in refractive power of the principal meridians.

Astigmatism of the fundus of the eye has been discussed by Siebeck (1960).

# Chapter Four
# Visual Optics of Emmetropia

## 4.1 Introduction

In this section a start will be made with visual optics. First the fictitious eyes used for calculations will be discussed: Gullstrand's schematic eyes were considered to be too complicated for this purpose. Instead, reduced eyes and a theoretical eye were proposed. Having discussed these the constants of the emmetropic standard reduced eye will be set out.

## 4.2 The concept of the reduced eye

As was shown in section 2.5 the optical system of the human eye is not a simple one. Visual optics calculations would, therefore, become very involved. To simplify matters a number of approaches were suggested, all along the same lines. The following quotation from Donders (1864) explains these simplifications:

'Following Listing's example (1853), we may go still a step further in the simplification: it is, in fact, allowable to reduce the compound dioptric system of the eye to a single refracting surface, bounded anteriorly by air, posteriorly by aqueous or vitreous humour, and this reduced eye, where the greatest accuracy is not required, may be made the basis of a number of considerations and calculations. With this simplification we can, with the greatest ease, form a satisfactory idea of the magnitude of the retinal images, of the position of the conjugate foci, of the extent of the circles of diffusion in imperfect accommodation, in astigmatism, etc., and of numerous other points.

The right to this simplification we derived from the minuteness of the distance between the two nodal points and between the two principal points of the dioptric system of the eye; this distance amounts to less than one-fourth of a millimetre. It is evident that neglecting this will

ause only a very slight difference. We thus obtain, besides the two ocal points, as cardinal points, only one principal point, and one nodal oint, the latter being the optical centre: that is, we retain simply the ardinal points of one simple refracting surface . . . '

Referring to *Figure 1.22* and the notes about the nodal points in secion 1.2, it can be shown that a system such as the eye, where the refracive indices on either side of the optical system differ, can be represented y a single refracting surface separating the two media of different refracive index (Jalie, 1977). In *Figure 4.1* it can be seen that ΔΔHPN and

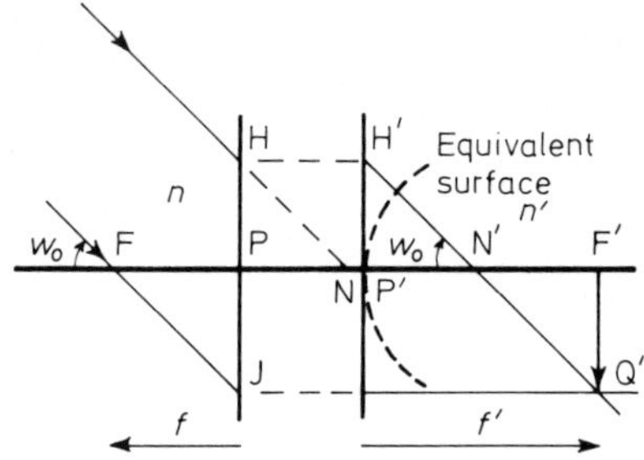


*igure 4.1 Determination of the osition and radius of curvature of he 'equivalent surface' of an optial system separating two media of ifferent refractive index (after alie, 1977)*

H′P′N′ are congruent. Hence PN = P′N′. Note that the nodal points re displaced towards the denser medium $n'$. ΔΔFPJ and N′F′Q′ are ongruent and FP = N′F′ = $-f$. Since P′F′ = $f'$, P′N′ = P′F′ − N′F′ = $f' - (-f) = f' + f$ which, according to equation (1.6), may be written as

$$\begin{aligned} P'N' &= \frac{n'}{F} + \frac{-n}{F} \\ &= \frac{n' - n}{F} \end{aligned}$$

This is a form of equation (1.4). Hence

$$P'N' = r$$

which proves the above thesis. Thus, an optical system separating two media of different refractive index may be replaced by a single refracting surface (the 'equivalent surface') whose vertex lies at P′, whose centre of curvature lies at N′, while its radius is given by

$$\begin{aligned} r &= P'N' \\ &= \frac{n' - n}{F} \end{aligned}$$

where $F$ is the equivalent power of the system.

The calculations in this text will be based on a reduced eye developed by Emsley based, in turn, on the reduced eye developed by Listing (1853). The following details were taken from the 1955 (5th) edition of his book. See also *Figures 4.2*, *4.3* and *4.4.*

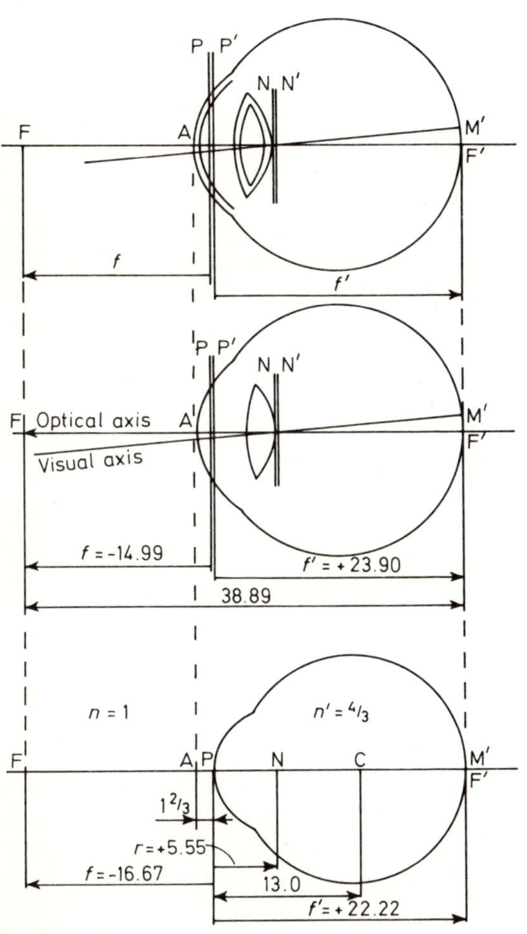


*Figure 4.2 Gullstrand's no.1 Exact Schematic Eye*
*Figure 4.3 Gullstrand's no.2 Simplified Schematic Eye*
*Figure 4.4 The Reduced Eye*

Emsley based his reduced eye on a modification of Gullstrand's no.2 Simplified Schematic Eye. Instead of a refractive index for the humours of 1.336 and for the lens of 1.413 (*see Table 2.1*), Emsley used the values 4/3 (water) and 1.416. He explained that this produced a reduced eye, derived from this schematic eye, of almost exactly +60 D power. On this basis he showed the no.2 emmetropic eye to be of +60.48 D power, with the anterior focal point very nearly at −15 mm. The principal points situated closely together are fairly close to the anterior corneal surface.

Emsley then calculated that the position of the first focal point is at −14.99 mm and that of the second at +23.90 mm (compare *Table 2.1* and *Figure 4.2*).

In a further discussion of his reduced eye he stated that it is desirable for the anterior and posterior focal points of the reduced eye and of the schematic eye to coincide. This fixed the power and, consequently, the radius of curvature and position of the single refracting surface.

Emsley's reduced eye's data may be derived as follows. The distance between the anterior and posterior focal points of Emsley's modification of the Gullstrand no.2 Schematic Eye is

$$FF' = 14.99 + 23.90 = 38.89 \text{ mm}$$

The focal lengths ratio is equal to that of the refractive indices used, that is, external refractive index (air) $n = 1$, and internal refractive index (water) $n' = 4/3$. In other words:

$$\frac{\text{posterior focal length}}{\text{anterior focal length}} = -\frac{4/3}{1} = -\frac{4}{3}$$

It follows that the anterior focal length

$$f = -\frac{3}{7}(38.89) = -16.67 \text{ mm}$$

and the posterior focal length

$$f' = \frac{4}{7}(38.89) = +22.22 \text{ mm}$$

The power of the eye is (equation (1.6))

$$F = \frac{-n}{f} = \frac{-1 \times 1000}{-16.67} \approx +60.00 \text{ D}$$

or, alternatively,

$$F = \frac{n'}{f'} = \frac{4/3\ (1000)}{+22.22} \approx +60.00\ \text{D}$$

Finally, the position of the single refracting surface or 'reduced surface' P with respect to the anterior corneal surface A of (Emsley's modification of) the Schematic Eye (*Figures 4.3* and *4.4*) is:

$$\text{AP} = \text{FP} - \text{FA} = 16.67 - 14.99 = 1.68\ \text{mm}$$

or, say, $1\frac{2}{3}$ mm.

The above can be summarised as follows:

The reduced eye consists of a single refracting surface of +60.00 D power which separates air ($n = 1$) from water ($n' = 4/3$) and which is situated $1\frac{2}{3}$ mm behind the anterior corneal surface of the Schematic Eye (*Figure 4.4*).

NOTE

The entrance pupil is situated in the plane of the reduced surface (*see* section 20.3).

NOTE

When the power of a reduced eye differs from the standard value of +60.00 D, the distance AP will differ too though only very slightly. Therefore, in calculations the distance AP is always assumed to be $1\frac{2}{3}$ mm

To quote Graves (1968) once more: all reasonable choices lead to neighbouring results. This is evident from *Table 4.1*. It shows the

TABLE 4.1

Comparison of Gullstrand's Exact Schematic Eye and the reduced eyes (measurements in millimetres)

| | *Gullstrand* | *Listing* | *Donders* | *Emsley* |
|---|---|---|---|---|
| Principal point | +1.348 | +2.344 8 | +2 | +1.67 |
| Radius of curvature | +5.7 | +5.124 8 | +5 | +5.55 |
| Anterior focal length | −17.055 | −15.177 4 | −15 | −16.67 |
| Posterior focal length | +22.785 | +22.474 2 | +20 | +22.22 |

values of Gullstrand's Exact Schematic Eye (no.1) and those of the reduced eyes of Listing, Donders and Emsley.

## 4.3 The concept of the theoretical eye

More recently Le Grand (first edition 1946; 1964) developed another fictitious eye, the 'theoretical eye'. It represents 'a reasonable mean of the adult European eye'. Le Grand used the following constants for the unaccommodated theoretical eye:

| | |
|---|---|
| Position: anterior lens surface | +3.6 mm |
| posterior lens surface | +7.6 mm |
| Radii of curvature: cornea | +8.0 mm |
| anterior lens surface | +10.2 mm |
| posterior lens surface | −6.0 mm |
| Refractive index for wavelength 589 nm at 18 °C: | |
| aqueous humour | 1.337 4 |
| lens | 1.42 |
| vitreous humour | 1.336 |

The following data may be calculated from these constants:

| | |
|---|---|
| Power: cornea | +42.36 D |
| lens | +21.78 D |
| total power of optical system | +59.94 D |

Ivanoff (1953) criticized the basis of the refractive indices used by Le Grand: they should not have been measured at a temperature of 18 °C, but at the normal body temperature of 37 °C. As for the varying refractive index of the lens, Ivanoff showed that a refractive index higher than 1.42 should be chosen. His modifications of the refractive indices of the theoretical eye, for wavelength 589 nm at 37 °C, were:

| | |
|---|---|
| Aqueous humour | 1.335 4 |
| Lens | 1.44 |
| Vitreous humour | 1.334 |

## 4.4 The standard reduced emmetropic eye

The standard reduced eye henceforth to be used for purposes of explanation and calculation, has the constants of Emsley's reduced eye. One further constant will be used, namely that of the axial length. The axial length $k'$ is the distance measured along the optical axis from the reduced surface P which could be said to act as cornea, to the macular area $M'$ of the retina. When the second principal focus $F'$ and the

macular area $M'$ (henceforth to be referred to as 'the retina') coincide, that is when $f' = k'$, the eye will be emmetropic (*Figure 4.4*).

It is possible to design any number of emmetropic reduced eyes; the only condition to be fulfilled is that $f' = k'$.

To indicate that the standard constants of the reduced eye are being used, the suffix 'o' will be added to the symbols. This will make it easy to recognize the source of any ametropia, as will be shown below. Moreover, it will not be necessary to restate the constants of the standard reduced eye time and again.

The reader is advised to familiarize himself with these constants (*Table 4.2*).

TABLE 4.2

The constants of the standard reduced emmetropic eye

| | | |
|---|---|---|
| Refractive power | $F_o$ | +60.00 D |
| External refractive index (air) | $n$ | 1 |
| Internal refractive index (water) | $n'$ | $4/3 = 1.333$ |
| Principal point position (AP) | | 1.67 mm |
| Derived constants: | | |
| radius of curvature | $r_o$ | +5.55 mm (equation (1.4)) |
| first focal length | $f_o$ | −16.67 mm (equation (1.8)) |
| second focal length | $f'_o$ | +22.22 mm (equation (1.9)) |
| axial length | $k'_o$ | +22.22 mm |

The diagrammatic representation of the reduced eye which will be used for simplicity of drawing, is shown in *Figure 4.5*.

NOTE

Since it is unusual for the refractive indices of the eye, and therefore of the reduced eye, to vary from the usual values, the suffix 'o' will be omitted. Thus, unless specifically stated otherwise, the refractive indices of any reduced eye will be $n = 1$ and $n' = 4/3$ or 1.333.

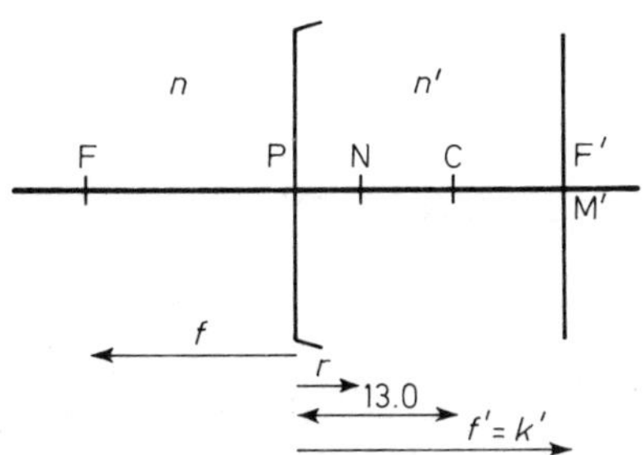


*Figure 4.5 Diagrammatic representation of the Reduced Eye. C represents the centre of projection (see section 18.1)*

## 4.5 The non-standard reduced eye

A reduced eye need not be emmetropic nor do its measurements have to conform to the standard constants. If a measurement is non-standard, its symbol will not be followed by the suffix 'o'; in the case of an eye of non-standard power the power symbol will be followed by the suffix 'e' indicating that this is the power of the 'eye'. For example, a reduced eye of +65.00 D would have the following measurements:

| | | |
|---|---|---|
| Power | $F_e = +65.00$ D | |
| First focal length | $f = \dfrac{-1 \times 1000}{+65.00}$ | $= -15.38$ mm |
| Second focal length | $f' = \dfrac{1.333 \times 1000}{+65.00}$ | $= +20.51$ mm |
| Radius of curvature | $r = \dfrac{(1.333 - 1)1000}{+65.00}$ | $= +5.12$ mm |

If this eye would have an axial length $k' = +20.51$ mm, which would mean that $f' = k'$, the condition for emmetropia would be fulfilled. On the other hand, if the axial length were standard, that is $f' \neq k'_o$, or if, say, $k' = +21.00$ mm, it would mean that $f' \neq k'$ and the eye would be ametropic.

## 4.6 The retinal image size

The retinal image size depends on the angle subtended by the object at the eye's principal point. The situation is comparable to that of *Figure 1.2*, though in the reduced eye it is the reduced surface which separates two media of different refractive index.

Whenever there is an image, there must be an object. One could assume two fundamentally different positions for the object: either it is at infinity or it is elsewhere. First the retinal image size of a distant object in the standard reduced emmetropic eye will be considered and this will be followed by a short note about the case where the object is not situated at infinity.

### 4.6.1 THE DISTANT OBJECT

Since emmetropia is defined with respect to infinity, the image of a distant object will be formed on the retina of the emmetropic eye.

The size of a distant object cannot be given in units of length, but only as an angular measurement. This usually takes the form of 'a distant object subtends an angle $w$ at the eye' (*Figure 4.6*). Since the object is

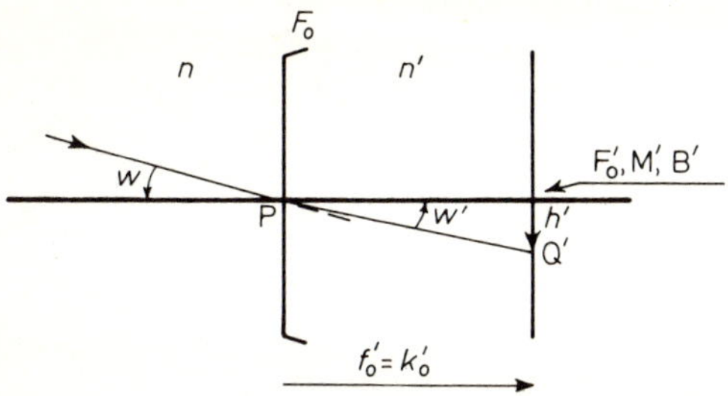


*Figure 4.6 Image formation of a distant object subtending an angle w at an emmetropic reduced eye*

distant, that is, situated at infinity, it is not necessary to specify whether the angle is measured at the principal point P or at any other point on the optical axis of the system of the eye because the angle does not vary in subtense.

Consider *Figure 4.6*. The angle $w$ subtended by the distant object is measured at the principal point P. At this point refraction takes place according to Snell's law (equation (1.1)). Thus

$$n \sin w = n' \sin w'$$

In visual optics only paraxial rays are considered and so the expression may be written as

$$nw = n'w'$$

or

$$w' = \frac{n}{n'} w = \frac{1}{4/3} w = \frac{3}{4} w$$

Now

$$\tan w' = \frac{M'Q'}{M'P} = \frac{-h'}{k'_o}$$

and so the retinal image size in the standard reduced emmetropic eye is given by

$$h' = -k'_o \tan w' \quad (4.1)$$

$$= -k'_o \frac{3}{4} \tan w$$

$$= -22.22 \times \frac{3}{4} \tan w$$

$$= -16.67 \tan w \text{ mm} \quad (4.2)$$

However, if the emmetropic eye has non-standard components, $k'_o$ must be substituted by $k'$ and so equation (4.1) becomes

$$h' = -k' \tan w' = -k' \frac{3}{4} \tan w \quad (4.3)$$

Since M′, B′ and $F'_o$ all indicate the same point (*Figure 4.6*) on the optical axis, two of the three symbols are often omitted, namely B′ and $F'_o$. B′ and Q′ do not give information that cannot be obtained from another source, for the arrow indicates the position and the size and whether or not the image is inverted.

Whenever there is a distant object the magnification formula of equation (1.15) which reads

$$m = \frac{L}{L'}$$

cannot be used since $L = 0$ which puts $m = 0$.

### 4.6.2 OBJECT NOT AT INFINITY

If the object is not situated at infinity the object vergence at the emmetropic eye will not be zero. The result will be that the image vergence will not be equal to the power of the eye and consequently $l' \neq k'_o$, that is, the image will not be formed on the retina. This case will be dealt with in chapter 6.

# Chapter Five
# The Ametropia Formula

## 5.1 Introduction

A reduced eye may be equated with a simple optical system consisting of an object space, a refracting system and an image space. For such a system the conjugate foci formula, equation (1.12)

$$L + F = L'$$

is used. However, in the eye there is a distinct difference in that any object should, ideally, be imaged on the retina. For this to occur the vergence inside the eye must depend on the axial length $k'$ and the internal refractive index $n'$. This vergence, henceforth to be called 'the dioptric length' (which is a convenient term although from an optical point of view, it is without foundation) is designated $K'$ and is given by

$$K' = \frac{n'}{k'} \tag{5.1}$$

It is measured at the reduced surface (*Figure 5.1*).

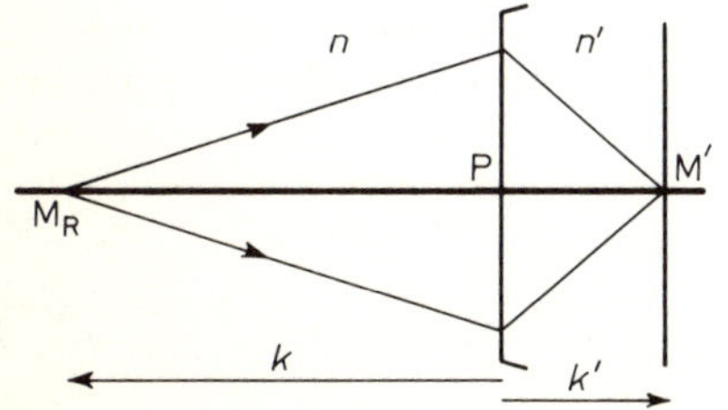


*Figure 5.1 The myopic eye and its far point* $M_R$

If the axial length is that of the standard reduced eye, the dioptric length is

$$K'_o = \frac{n'}{k'_o} = \frac{1.333 \times 1000}{+22.22} \approx +60.00 \text{ D}$$

## 5.2 Application

The power of the reduced eye will be either standard $F_o$, or non-standard $F_e$. The same applies to the dioptric length $K'$. The ocular correction $K$, also called the ocular refraction, the principal point refraction, ocular ametropia or, simply, ametropia, is then given by the following modification of the conjugate foci formula, to be called 'the ametropia formula',

$$K = K' - F_e \tag{5.2}$$

This general expression may be used to indicate the following types of emmetropia or ametropia when appropriate suffixes are added:

Emmetropia ($K = 0$) in the standard reduced eye:

$$K = K'_o - F_o = 0 \tag{5.3}$$

Refractive ametropia in the reduced eye:

$$K = K'_o - F_e \tag{5.4}$$

Axial ametropia in the reduced eye:

$$K = K' - F_o \tag{5.5}$$

Emmetropia ($K = 0$) or ametropia ($K \neq 0$) in the non-standard reduced eye:

$$K = K' - F_e \tag{5.6}$$

The ocular correction $K$ is measured at the reduced surface (*Figure 5.1*). In the same way as the dioptric length is comparable to the image vergence, the ocular correction is comparable to the object vergence of the conjugate foci formula. It follows (equation (1.10)) that

$$K = \frac{n}{k} \tag{5.7}$$

where $k$ is the 'far point distance' which is measured from the principal point P to the far point $M_R$. The far point is the point on the optical axis which is conjugate with the retina $M'$. This point is also known as punctum remotum. The far point plane is perpendicular to the optical axis at the far point $M_R$, and it is conjugate with the retina.

Since the eye is usually situated in air, equation (5.7) usually reduces

to $K = 1/k$. However, this is not the case when the eye is, for instance, immersed in water during swimming!

In case of emmetropia the far point will be at infinity. The far point distance is then

$$k = \infty$$

and the ocular correction

$$K = 0 \text{ D}$$

Thus, in emmetropia, $K = 0$ D as indicated at the beginning of this section.

Examples of the use of the ametropia formula are given in the next chapter.

NOTE

The conjugate foci formula is a general formula which may be applied in all cases. However, the ametropia formula is, in principle, only meant to be used to calculate whether a reduced eye is emmetropic or ametropic. If the eye is ametropic, the appropriate formula will also furnish the type of ametropia and its amount. If an object is situated in the far point plane of an eye, the ametropia formula and formulae derived thereof, may be used to calculate the position of the retina and the retinal image size. On the other hand, if the object is not situated in the far point plane the conjugate foci formula will have to be used to find the position of the image plane and the image size, with the prior knowledge that image plane and retina do not coincide.

The conjugate foci formula and the ametropia formula serve different purposes, and they should not be confused.

# Chapter Six
# Visual Optics of Spherical Ametropia

## 6.1 Introduction

In section 3.3.1 spherical ametropia was classified into two main divisions:

1. Axial
2. Refractive

In section 3.4.1 it was also classified into:

1. Myopia
2. Hyperopia

In the following sections the combination of the two classifications will be dealt with. For this purpose it will be assumed that, if a reduced eye is axially ametropic, it will be of standard refractive power, and if a reduced eye is refractively ametropic, it will be of standard axial length.

## 6.2 Myopia

### 6.2.1 AXIAL MYOPIA

In myopia a distant object will be imaged in front of the eye's retina (*Figure 6.1*). This means that the axial length is longer than the eye's second focal length. Thus in axial myopia

$$k' > f'_o$$

Consequently,

$$\frac{n'}{K'} > \frac{n'}{F_o}$$

or

$$F_o > K'$$

Consider a case of axial myopia where the axial length is $k' = +23.00$ m
What can be learned from this information?

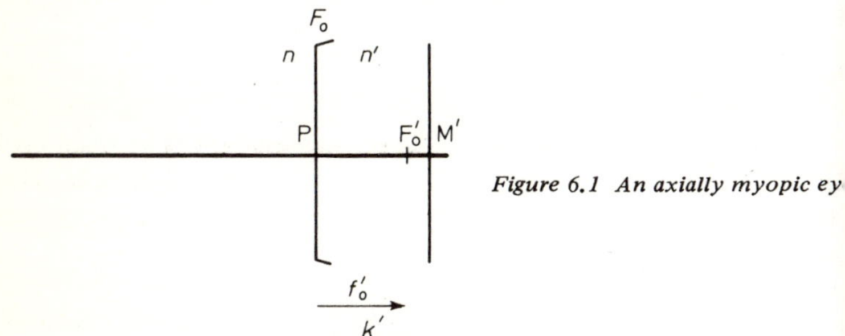


*Figure 6.1 An axially myopic ey*

Firstly, since this is a case of axial ametropia, the power of the eye is standard, and so is its focal length (*Figure 6.1*).

Secondly, the dioptric length may be calculated from equation (5.1), namely

$$K' = \frac{1.333 \times 1000}{+23.00} = +57.96 \text{ D}$$

Thirdly, this information may be substituted into the ametropia formul (equation (5.5)),

$$K = K' - F_o$$

$$= +57.96 - (+60.00)$$

and so the ocular correction is $K = -2.04$ D.

Fourthly, from equation (5.7) the far point distance may be calculated:

$$k = \frac{1 \times 100}{-2.04} = -49.02 \text{ cm}$$

Thus, if an object were situated 49.02 cm in front of the reduced surface of this eye, it would be imaged on the retina because the far point plane and retina are conjugate planes (*Figure 6.2*).

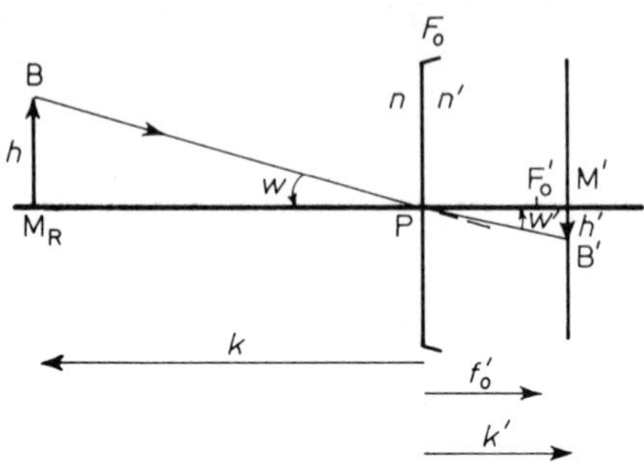


*Figure 6.2 The image formation of an object situated at the far point of an axially myopic eye*

This, too, explains the term shortsightedness. An object situated as near as half a metre from the eye, can be seen clearly. According to the clinical classification, an ocular correction of −2.04 D is only a low degree of myopia. Myopia is considered to be low up to an ocular correction of −3.00 D which means a far point distance of −33.33 cm. This shows that the higher the degree of myopia, the shorter the far point distance.

From these data the magnification of the system may be calculated (*Figure 6.2*). There are two approaches:

1. The geometrical approach:
   Consider $\Delta\Delta M_R BP$ and $M'B'P$:

$$\tan w = \frac{h}{-k} \quad \text{and} \quad \tan w' = \frac{-h'}{k'}$$

and in view of equation (4.3),

$$\tan w' = \frac{k' \times \frac{3}{4} \tan w}{k'} = \frac{3}{4} \tan w \qquad (6.1)$$

or

$$\frac{-h'}{k'} = \frac{3}{4} \times \frac{h}{-k}$$

which may be rearranged to

$$\frac{h'}{h} = \frac{3}{4} \times \frac{k'}{k}$$

In view of equation (1.13), and by substituting the values shown above, the magnification is found to be

$$m = \frac{3}{4} \times \frac{k'}{k} \qquad (6.2)$$

$$= \frac{3 \times 23.00}{4(-490.2)} = -0.035\,2$$

This means that the retinal image size is 0.035 2 times the object size.

2. The use of the magnification equation (equation (1.15)) which reads

$$m = \frac{L}{L'}$$

Since the ametropia formula is a particular form of the conjugate foci formula, where $L$ is replaced by $K$ and $L'$ by $K'$, the magnification equation may also be written as

$$m = \frac{K}{K'} \qquad (6.3)$$

and applied where appropriate, that is, only when the object lies in the far point plane of the eye.
In this example the magnification calculated by means of equation (6.3), is found to be

$$m = \frac{-2.04}{+57.96} = -0.035\,2$$

It will be seen that the second approach involves fewer calculations; it is, therefore, quicker and offers fewer opportunities for calculational errors. For these reasons it is suggested that the latter approach should be chosen when possible.

If the object size MB $= h = +0.5$ cm, the retinal image size M$'$B$'$ $= h' = mh = -0.035\,2 \times 0.5$ (cm) $= -0.035\,2 \times 5$ (mm) $= -0.18$ mm. The retinal image size is given in millimetres since it is a measurement within the eye.

NOTE
It will be found that retinal image sizes seldom exceed 0.5 mm when the image is formed on the retina.

NOTE
When the magnification of optical systems associated with the eye is calculated to four decimal places and rounded off to two decimal places,

the retinal image sizes measured in millimetres will agree very closely with those calculated on the basis of the geometry of the optical pathways.

### 6.2.2 REFRACTIVE MYOPIA

In myopia the image of a distant object is formed in front of the retina of the eye. In refractive myopia this is due to the standard axial length

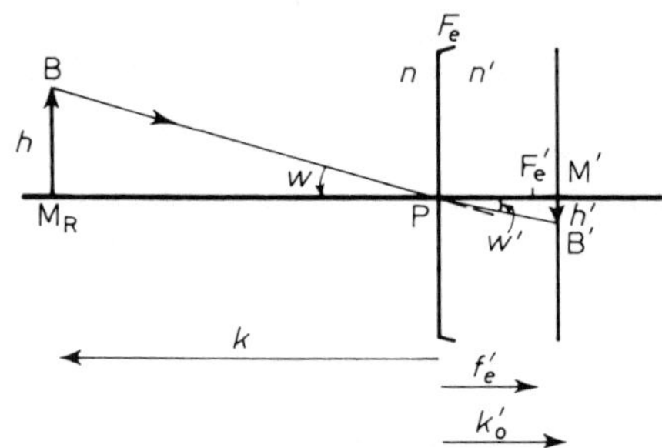


*Figure 6.3 The image formation of an object situated at the far point of a refractively myopic eye*

$k_o'$ being longer than the second focal length $f_e'$ (*Figure 6.3*). This can be indicated by

$$k_o' > f_e'$$

Consequently,

$$\frac{n'}{K_o'} > \frac{n'}{F_e}$$

or

$$F_e > K_o'$$

Assuming a similar degree of myopia as in section 6.2.1, say $K = -2.00$ D, and substituting this into the ametropia formula, equation (5.4), the power of the eye is found to be

$$F_e = K_o' - K$$

$$= +60.00 - (-2.00)$$

$$= +62.00 \text{ D}$$

These data form the basis for deriving the following data for this eye:

Far point distance (equation (5.7))

$$k = \frac{1 \times 100}{-2.00} = -50.00 \text{ cm}$$

Second focal length (equation (1.9))

$$f'_e = \frac{1.333 \times 1000}{+62.00} = +21.50 \text{ mm}$$

The magnification of the optical system may be calculated in either of the following ways:

1. The geometrical approach:
   Consider $\Delta\Delta M_R BP$ and $M'B'P$; equation (6.2), now becomes

$$m = \frac{3}{4} \times \frac{k'_o}{k}$$

$$= \frac{3 \times 22.22}{4 \times -500} = -0.033\,3$$

2. Using the magnification formula, equation (6.3) becomes

$$m = \frac{K}{K'_o} = \frac{-2.00}{+60.00} = -0.033\,3$$

If the object size is once again $h = +0.5$ cm or $+5$ mm, the retinal image size $h' = mh = -0.033\,3 \times 5 = -0.17$ mm.

Compare the retinal image size obtained here with that of the previous and following sections.

## 6.3 Hyperopia

### 6.3.1 AXIAL HYPEROPIA

Assuming no accommodation (chapter 12) takes place, a distant object will be imaged behind the retina of a hyperopic eye. This means that the eye's second focal length is longer than its axial length. In axial hyperopia this may be indicated as

$$f'_o > k'$$

Consequently,

$$\frac{n'}{F_o} > \frac{n'}{K'}$$

or

$$K' > F_o$$

Consider a case of axial hyperopia where the axial length $k' = +21.75$ mm. Hence:

Firstly, the power and length of the eye are those of the standard reduced eye since this is a case of axial hyperopia.

Secondly, the dioptric length may be calculated from equation (5.1), namely

$$K' = \frac{1.333 \times 1000}{+21.75} = +61.29 \text{ D}$$

Thirdly, the ocular correction is found by substituting these values into the ametropia formula (equation (5.5)),

$$\begin{aligned} K &= K' - F_o \\ &= +61.29 - (+60.00) \\ &= +1.29 \text{ D} \end{aligned}$$

Fourthly, the far point distance may be calculated from equation (5.7):

$$k = \frac{1 \times 100}{+1.29} = +77.52 \text{ cm}$$

Thus, if an object were situated 77.52 cm behind the reduced surface it would be imaged on the retina since the far point and retina are conjugate (*Figure 6.4*).

On first sight this sounds rather strange: the object must be placed behind the eye for it to be imaged on the retina of that eye! However, the concept of the virtual object and virtual image has been discussed (section 1.2). The object at the far point of a hyperopic eye is a case in point. Therefore, if it can be arranged by some optical means for a (virtual) image of a real object (situated in front of the eye) to be formed in the far point plane (behind the eye), then this virtual image will act

as if it were a virtual object with respect to the eye. Thus, it will be imaged on the retina of the hyperopic eye. The reader will appreciate that the 'optical means' referred to above, will be a spectacle lens.

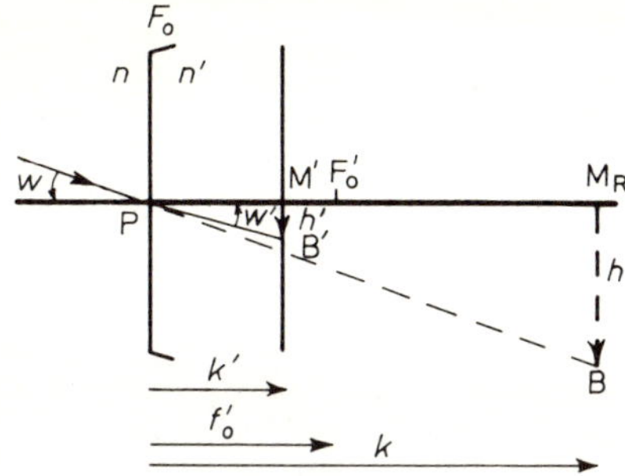


*Figure 6.4 The image formation of a virtual object situated at the far point of an axially hyperopic eye*

Another line of approach is possible in this context. The reader will have noticed that myopia and hyperopia are opponents: where myopia has a negative ocular correction, that of the hyperopic eye is positive; where the myope's far point distance is negative, that of the hyperopic eye is positive, and this antagonism is encountered throughout. When the emmetropic far point is approached from the high myope's far point, the far point moves from a position close to the eye through the 'negative' zone in front of the eye, until infinity is reached. Infinity, however, is not only situated in the 'negative' zone, to the left of the eye, but also on the far right, in the 'positive' zone. If the emmetropic far point is approached from that of a high hyperope, the far point position will be seen to move from closely behind the eye to infinity to the right of the eye, through the 'positive' zone. For example, if an eye had an ocular correction of $K = +10.00$ D, the far point would be at +10.00 cm, that is, at 10 cm behind the reduced surface. And so with respect to the direction in which the eye is looking, the hyperopic far point is very far away, 'on the other side' of infinity. Hence, the name farsightedness.

Returning to the above example, the retinal image size may be calculated as above:

1. The geometrical approach:
   Consider $\Delta\Delta M_R BP$ and $M'B'P$; equation (6.2) then becomes

$$m = \frac{3}{4} \times \frac{k'}{k}$$

$$= \frac{3 \times (+21.75)}{4 \times (+775.2)} = +0.021\,0$$

2. Using the magnification equation, equation (6.3) becomes

$$m = \frac{K}{K'} = \frac{+1.29}{+61.29} = +0.021\,0$$

Assuming the same object size of 5 mm, the retinal image size will be given by equation (1.13):

$$h' = mh = (+0.021\,0) \times 5 = +0.11 \text{ mm}$$

### 6.3.2 REFRACTIVE HYPEROPIA

In this case the power of the eye is non-standard, but its axial length is standard. Therefore,

$$f'_e > k'_o$$

Consequently,

$$\frac{n'}{F_e} > \frac{n'}{K'_o}$$

or

$$K'_o > F_e$$

If an ocular correction of $K = +1.25$ D is assumed, the power of the ametropic eye is given by equation (5.4)

$$F_e = K'_o - K$$

$$= +60.00 - (+1.25)$$

$$= +58.75 \text{ D}$$

The second focal length is given by equation (1.9)

$$f'_e = \frac{n'}{F_e} = \frac{1.333 \times 1000}{+58.75} = +22.69 \text{ mm}$$

and the far point distance by equation (5.7)

$$k = \frac{n}{K} = \frac{1 \times 100}{+1.25} = +80.00 \text{ cm}$$

(See *Figure 6.5*.)

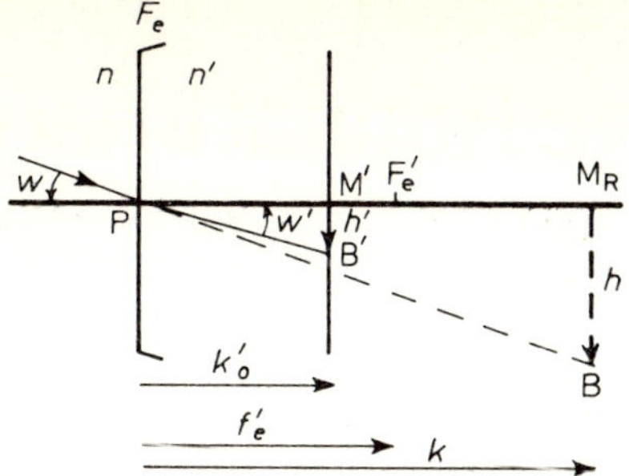


*Figure 6.5 The image formation of a virtual object situated at the far point of a refractively hyperopic eye*

The magnification of the system ($m$) may be found using either equation (6.2) or (6.3). From equation (6.2)

$$m = \frac{3}{4} \times \frac{k'_o}{k}$$

$$= \frac{3 \times (+22.22)}{4 \times (+800.0)}$$

$$= +0.020\,8$$

From equation (6.3),

$$m = \frac{K}{K'_o}$$

$$= \frac{+1.25}{+60.00}$$

$$= +0.020\,8$$

For an object of 5 mm this would give a retinal image size of (equation (1.13))

$$h' = mh = (+0.020\,8) \times 5 = +0.10 \text{ mm}$$

# Chapter Seven
# The Basic Retinal Image Size

## 7.1 Introduction

The basic retinal image size may be defined as the retinal image whose size is determined by the chief ray, that is, the ray in a pencil of light which passes through the centre of the entrance pupil of an optical system. In the case of the reduced eye the entrance pupil and the reduced surface are considered to coincide (*see* section 4.2).

From the geometry of the reduced eye, the relative sizes of the basic retinal images of different types of eyes, may be determined.

## 7.2 Refractive ametropia

Consider *Figure 7.1.* Although the power of refractively ametropic reduced eyes vary, their axial length is the same. Since the power of the

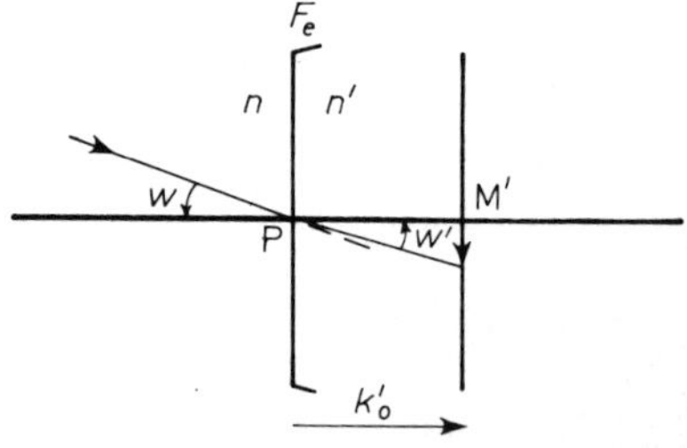


*Figure 7.1 The basic retinal image formation in a refractively ametropic eye*

eye does not affect the relationship between the incident and refracted chief ray, the basic retinal image size is the same in all such eyes and depends on the angle of incidence $w$ and the axial length. In this context it should not be overlooked that an emmetropic eye need not have an axial length of 22.22 mm.

## 7.3 Axial ametropia

Consider *Figure 7.2.* The incident and refracted chief ray have again the same relationship. However, the basic retinal images are formed at different distances from the reduced surface because of the varying

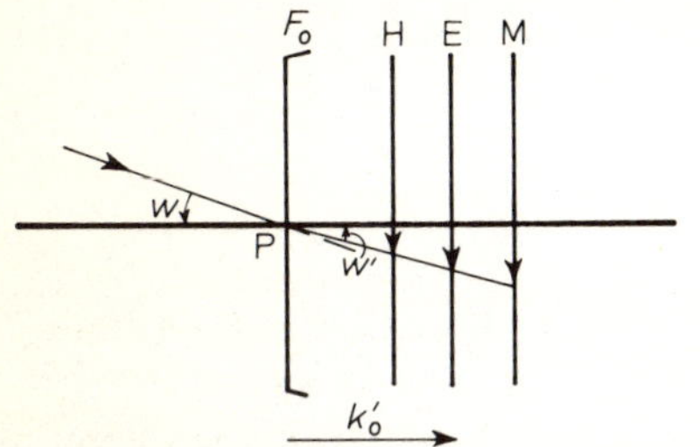


*Figure 7.2 The basic retinal image formation in axially ametropic eyes*

axial lengths of these eyes. Consequently, the basic retinal image is largest in the myopic eye and smallest in the hyperopic eye. Or:

$$h'_{\text{hyperope}} < h'_{\text{emmetrope}} < h'_{\text{myope}}$$

where $h'$ indicates the basic retinal image size.

For the calculation of the retinal image size the axial length of each eye needs to be known. It should be noted that these eyes need not be of standard power.

# Chapter Eight
# The Blurred Retinal Image in Spherical Ametropia

## 8.1 Introduction

In chapter 6 the optics associated with an object situated at the far point of a spherically ametropic eye was studied. This situation may occur from time to time, but more frequently the object will be situated elsewhere. Such situations will be dealt with in this chapter.

## 8.2 Point object

The real object may be situated anywhere in front of the eye, but on this occasion not at the far point of the eye; thus the ametropia formula cannot be applied, but the conjugate foci formula (equation (1.12)) and information derived thereof, does apply.

Since the object is not situated in the far point plane, it will not be imaged on the retina, but in front of or behind the retina. Consider the following case.

EXAMPLE 3
A distant point object is situated in front of a reduced eye. The axial length of the eye is 21.50 mm. The power of the eye is +59.46 D. Calculate the retinal image size when the diameter of the pupil is 5 mm.

*Solution 3*
*Step 1.* Draw a diagram on the basis of the data provided (*Figure 8.1*).

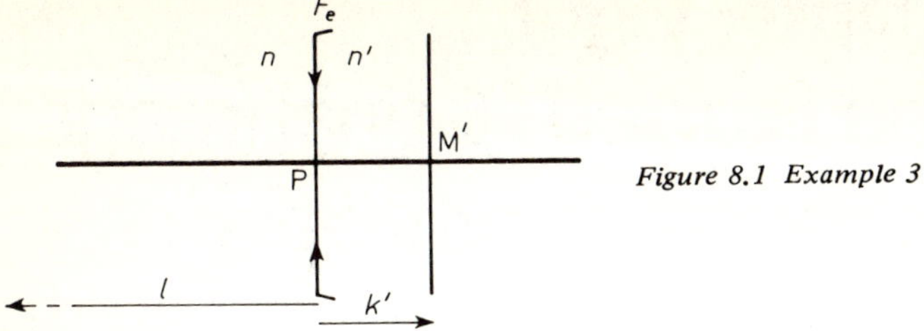


*Figure 8.1 Example 3*

*Step 2.* The object vergence may be found from the object distance. Thus:

$$l = \infty$$

and so

$$L = 0.00 \text{ D}$$

*Step 3.* The conjugate foci formula where the power of the optical system is $F_e$, provides the image vergence:

$$L' = L + F_e$$

$$= 0.00 + 59.46$$

$$= +59.46 \text{ D}$$

and the image distance will be

$$l' = \frac{1.333 \times 1000}{+59.46}$$

$$= +22.42 \text{ mm}$$

NOTE
As the axial length of this eye is $k' = +21.50$ mm, the image would be formed behind the retina if this were possible. In such cases the imaginary rays continuing beyond the retina are indicated by interrupted lines (*Figure 8.2*).

Since the distant object is imaged behind the retina, the eye must be hyperopic. The image will be a point because the object is a point. In such cases it is usually assumed that the object is situated on the optical

axis of the system. Consequently, the image (at B′) will be situated on that axis (*Figure 8.2*).

Only those rays passing through the pupil of the eye, the diameter of which is indicated by the two arrow heads at a and b in the reduced surface, can contribute to the image formation. This pencil is called the 'image forming pencil'. Rays reaching the eye from the object, but not passing through the pupil, are not normally indicated in diagrams unless they are used for optical construction purposes.

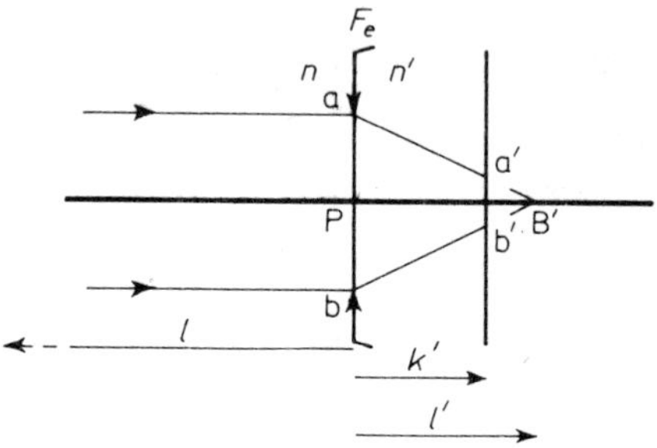

*Figure 8.2 Example 3; blur circle formation*

After refraction the image forming pencil takes the shape of the cone abB′ which is intersected by the retina. From the diagram it can be seen that the image forming pencil inside the eye forms a circle a′b′ on the retina. This circle is known as the blur circle or circle of diffusion.

The pupil diameter ab is denoted by $g$; the blur circle diameter a′b′ by $j$.

NOTE

In order to be able to calculate the blur circle diameter, the pupil diameter must be known. Consequently, whenever a pupil diameter is given amongst the data of a problem, one can be fairly certain that the blur circle diameter features somewhere in the solution of that problem!

*Step 4.* The blur circle diameter may be calculated as follows. Consider ΔΔabB′ and a′b′B′ (*Figure 8.2*):

$$\frac{j}{g} = \left(\frac{a'b'}{ab}\right)$$

$$= \frac{l' - k'}{l'}$$

or

$$j = \left|\frac{g(l' - k')}{l'}\right| \qquad (8.1)$$

Thus, by substitution, the blur circle diameter is found to be

$$j = \frac{5(22.42 - 21.50)}{22.42}$$

$$= \frac{5 \times 0.92}{22.42}$$

$$= 0.20 \text{ mm}$$

NOTE

The signs of the distances have been omitted in this formula since it is a simple geometrical and not an optical relationship.

It is common practice in optics to manipulate derivations so that vergences may be substituted for distances. Thus, equation (8.1) may be treated as follows:

$$j = g \times \frac{\frac{n'}{L'} - \frac{n'}{K'}}{\frac{n'}{L'}}$$

$$= g\left(\frac{n'}{L'} - \frac{n'}{K'}\right)\frac{L'}{n'}$$

$$= g\left(\frac{L'}{L'} - \frac{L'}{K'}\right)$$

$$= g\left(\frac{K'L' - L'^2}{K'L'}\right)$$

$$= \left|g\left(\frac{K' - L'}{K'}\right)\right| \qquad (8.2)$$

In order to be able to solve example 3 by substitution into this formula, the dioptric length of the eye needs to be calculated in addition to the data found in the above solution. Thus:

$$K' = \frac{1.333 \times 1000}{+21.50} = +62.00 \text{ D}$$

Substituting into equation (8.2):

$$j = 5\left(\frac{+62.00 - (+59.46)}{+62.00}\right) = 0.20 \text{ mm}$$

It is the author's opinion that for the better understanding of the actual situation equation (8.2) is better not used. In fact, since it is quite simple to determine from the similar triangles the relationship between pupil diameter and blur circle diameter, there is no need to memorize any formula at all.

Now consider a case of myopia. As an added complication the object will be neither at infinity nor at the eye's far point.

EXAMPLE 4

A reduced eye of −8.00 D ocular correction looks at a point object placed 4 m in front of it. Calculate the retinal image size when the pupil diameter is 7 mm and the ametropia is axial.

*Solution 4*

*Step 1.* Draw a diagram based on the data provided (*Figure 8.3*).

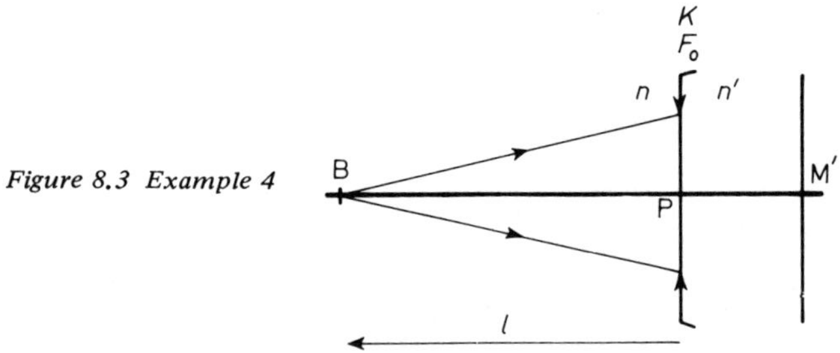


*Figure 8.3 Example 4*

*Step 2.* This is a case of axial ametropia, thus the power of the eye is $F_o$.
*Step 3.* Since the ocular correction and the power of the eye are known, the axial length may be calculated from the ametropia formula:

$$\begin{aligned} K' &= K + F_o \\ &= -8.00 + 60.00 \\ &= +52.00 \text{ D} \end{aligned}$$

Hence, the axial length is

$$\begin{aligned} k' &= \frac{1.333 \times 1000}{+52.00} \\ &= +25.63 \text{ mm} \end{aligned}$$

*Step 4.* Using the conjugate foci formula, the position of the point image may be calculated, thus:

Since $l = -4$ m, the object vergence $L = -0.25$ D. On substitution, the image vergence is

$$\begin{aligned} L' &= L + F_o \\ &= -0.25 + 60.00 \\ &= +59.75 \text{ D} \end{aligned}$$

and the image distance is

$$\begin{aligned} l' &= \frac{1.333 \times 1000}{+59.75} \\ &= +22.31 \text{ mm} \end{aligned}$$

*Step 5.* The image distance may now be shown in the diagram (*Figure 8.4*), and the image forming pencil may be drawn.

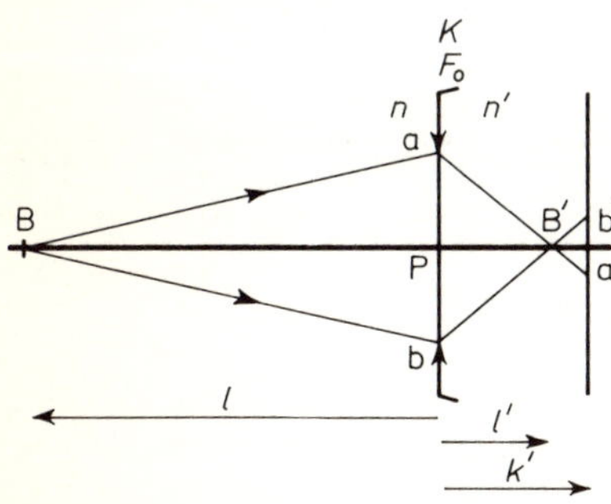


*Figure 8.4 Example 4; blur circle formation*

*Step 6.* From the similar triangles abB′ and a′b′B′, the blur circle diameter may be calculated:

$$\frac{j}{g} = \frac{k' - l'}{l'}$$

or

$$j = \left| g\,\frac{k' - l'}{l'} \right|$$

and by substitution, the blur circle diameter is

$$j = 7 \times \frac{25.63 - 22.31}{22.31}$$

$$= 1.04 \text{ mm}$$

Alternatively, using equation (8.2):

$$j = \left| g\frac{K' - L'}{K'} \right|$$

$$= 7 \times \frac{+52.00 - 59.75}{+52.00}$$

$$= -1.04 \text{ mm}$$

## 8.3 Extended object

In the previous section point objects and their blurred retinal images were discussed. In this section distant objects subtending a given angle and objects of a given size at a given distance, will be considered. The examples will be based on the data of examples 3 and 4 in the previous section.

EXAMPLE 5
A distant object subtends an angle $w$ at a reduced eye. The axial length of the eye is 21.50 mm and the power of the eye is +59.46 D. Calculate the retinal image size when the pupil diameter is 5 mm.

*Solution 5*
The following data were calculated in example 3:

$$l' = +22.42 \text{ mm}$$
$$j = 0.20 \text{ mm}$$

*Figure 8.5* shows the diagram based on the data provided. Any object may be considered as consisting of a very large number of 'point objects'. Following this approach, each point of the object may be dealt with separately, and afterwards the overall image may be built up on the basis of the separate details of the image.

In the previous section the point of the object situated on the optical axis was dealt with; its image position and the blur circle diameter it

formed on the retina were determined. Now, the other extremity of the object, the arrow head, will be considered.

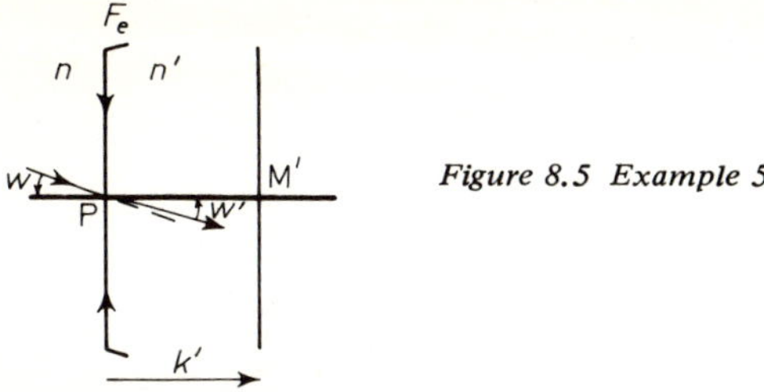


*Figure 8.5 Example 5*

*Step 1.* As stated above, the object will be imaged at a distance of $l' = +22.42$ mm from the reduced surface at P. Since this image lies beyond the retina, that portion of the image formation taking place 'virtually', is shown by an interrupted line (*Figure 8.6*).
*Step 2.* The size of the virtual image is of no real interest. From the geometry of the diagram (*Figure 8.6*) it will be seen that the basic retinal image B″Q″ = $h'$ is given by (equation (4.3))

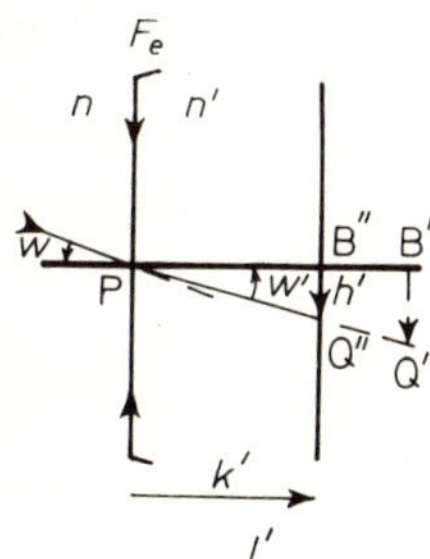


*Figure 8.6 Example 5; the position and size of the virtual image and the basic retinal image*

$$h' = -\frac{3}{4}\,k'\tan w$$

Thus, the basic retinal image size is

$$h' = -\frac{3}{4} \times 21.50 \tan w = -16.13 \tan w \text{ mm}$$

*Step 3.* This, however, does not take into account the fact that this retinal image is blurred. The calculation of the blur circle diameter of the image of one object point has been discussed. Now, this method will be applied to another object point: that of the arrow head.

Since the object point is at infinity all rays reaching the eye will be parallel. The extreme rays of the image forming pencil will pass within the pupil, and will focus at Q′ after refraction. This pencil is limited by the rays a and b (*Figure 8.7*). The blur circle thus formed on the retina around the arrow head of the basic retinal image is of size a′b′.

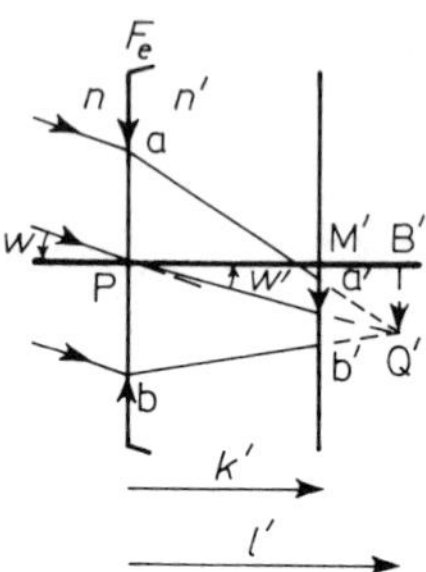


*Figure 8.7 Example 5; the blur circle formation around the chief ray* PQ′ *forming the retinal image of the arrow head*

*Step 4.* In *Figure 8.8*, the *Figures 8.6* and *8.7* have been combined. Although only the two blur circles associated with the extremities of the object are shown, it will be understood that every point of the basic retinal image is surrounded by such a blur circle (*see Figure 8.8*, inset).

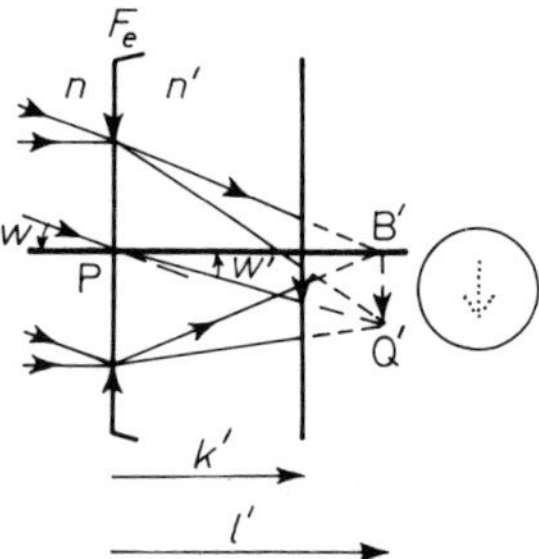


*Figure 8.8 Example 5; the blur circle formation around the retinal image points* B′ *and* Q′. *Inset: the blur circles for a number of retinal image points forming the image of the arrow*

The total retinal image size can be seen to consist of the basic retinal image, plus half a blur circle protruding beyond the foot, and half a blur circle protruding beyond the head of the image of the arrow. The total retinal image size is given by

$$|h'| + |j|$$

In this calculation the angle subtended by the object is only given in symbolic form. To obtain some idea of a commonly encountered size,

assume for $w$ a value of $1°45'$. Substituting this into equation (4.3), the basic retinal image size is

$$h' = -\frac{3}{4} \times 21.50 \times 0.030\,6 = -0.49 \text{ mm}$$

*Step 5.* Thus, the total retinal image size is

$$|h'| + |j| = 0.49 + 0.20 = 0.69 \text{ mm}$$

Next, the case of an object of given size at a given distance will be discussed. For this purpose, the details of example 4 (section 8.2) will be utilized.

EXAMPLE 6

A reduced eye of −8.00 D ocular correction has a 5 cm high object placed 4 m in front of it. Calculate the retinal image size when the pupil diameter is 7 mm and the ametropia is axial.

*Solution 6*

The following data were obtained in solution 4:

$$k' = +25.63 \text{ mm}$$
$$l' = +22.31 \text{ mm}$$
$$j = 1.04 \text{ mm}$$

*Figure 8.9* shows the diagram based on these data.

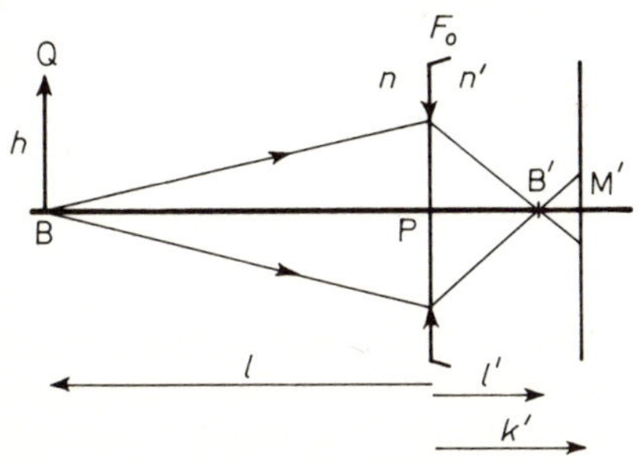


*Figure 8.9 Example 6; blur circle formation*

*Step 1.* The angle $w$ subtended by the object at the eye is given by

$$\tan w = \frac{h}{l}$$

*Step 2.* The angle $w'$ subtended by the basic retinal image, is given by (equation (6.1))

$$\tan w' = \frac{3}{4}\tan w$$

and so

$$h' = \frac{3 \times k' \times h}{4l}$$

*Step 3.* By substitution, the basic retinal image size $h'$ is

$$h' = \frac{3 \times 25.63 \times 50}{4 \times (-4000)}$$

$$= -0.24 \text{ mm}$$

*Step 4.* The total retinal image size is

$$|h'| + |j| = 0.24 + 1.04 = 1.28 \text{ mm}$$

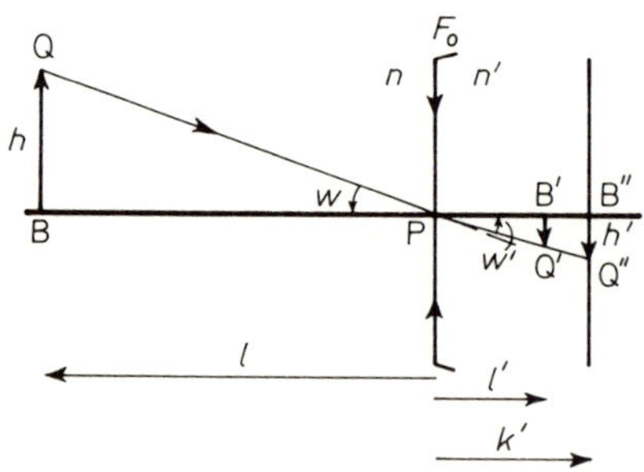


*Figure 8.10 Example 6; the formation and size of the image* B′Q′ *and the basic retinal image* B″Q″

The diagrams showing the retinal image formation are *Figures 8.9* and *8.10*. These could be combined into a single diagram showing both blur circle diameter and basic retinal image size.

## 8.4 Nomogram

*Figure 8.11* is a nomogram which may be used to find the ratio of the blur circle diameter and the pupil diameter, that is, $j/g$. The blur circle

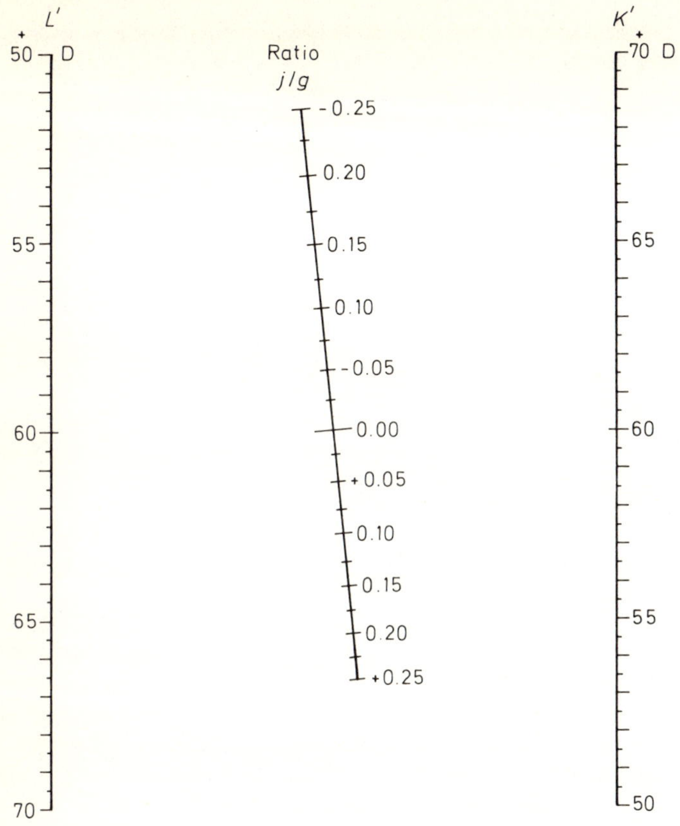


*Figure 8.11 Nomogram providing the ratio of blur circle diameter j and pupil diameter g*

diameter may be calculated by multiplying the pupil diameter and this ratio.

Taking example 3 (section 8.2) as example, the data required to use the nomogram are the image vergence and the dioptric length of the eye. In this case, $L' = +59.46$ D and $K' = +62.00$ D. Place a 'straight-edge', such as a ruler or the edge of a sheet of paper, across the $L'$-scale at +59.46 and across the $K'$-scale at +62.00. The ratio $j/g$ is now read off the centre scale, $j/g = -0.04$. The pupil diameter was given as 5 mm; the blur circle diameter is, therefore

$$j = 5 \times (-0.04) = -0.20 \text{ mm}$$

which agrees with the calculated value.

The details required from example 4 (section 8.2), are

$$L' = +59.75 \text{ D}$$
$$K' = +52.00 \text{ D}$$
$$g = 7 \text{ mm}$$

From the nomogram,

$$j/g = +0.15$$

The blur circle diameter is

$$j = 7 \times (+0.15) = +1.05 \text{ mm}$$

By calculation, the blur circle diameter was found to be 1.04 mm. If it had been possible to reproduce the nomogram on a larger scale a more accurate result would have been found.

NOTE
The nomogram is only meant as a check on calculated values. It will only be possible to obtain an estimate of the real value.

The positive and negative values of the ratio are associated with the position of the sharp image of the object and the retina. If the sharp image is formed in front of the retina, the ratio assumes a negative value and when the sharp image is formed behind the retina, it assumes a positive value (Obstfeld, 1975). Note that neither pupil diameter nor blur circle diameter carries a sign, and that the sign is given to the ratio only.

## 8.5 Apparent size

### 8.5.1 INTRODUCTION AND DISCUSSION

In visual optics the term apparent size refers to the size an object appears to have in space (the perceived size). It used to be determined by projecting the total retinal image through the nodal point to the object plane (*Figure 8.12*), as may be inferred from the representation of the horopter by early writers (Ogle, 1964). From a purely optical point of view, however, it is projected through the centre of the entrance pupil, that is the principal point of the reduced eye, thus using the chief ray (*Figure 8.13*). This method is subject to Snell's law. Either method

is suitable to indicate the inversion of the retinal image by the brain. It provides a pseudo-scientific explanation of the phenomenon whereby objects imaged on the inferior part of the retina are perceived as being situated in space above the visual axis (which is their true position: the hand can touch them there), or, if imaged on the nasal retina, they are perceived, and are situated on, the temporal side of the visual axis in space.

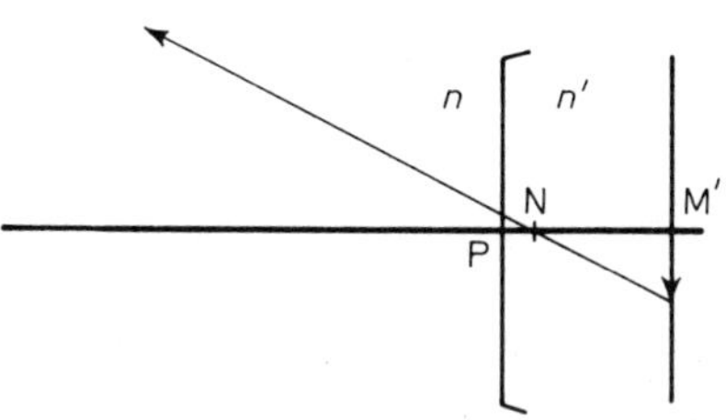


*Figure 8.12 Determination of the apparent size of an object by projection through the nodal point* N

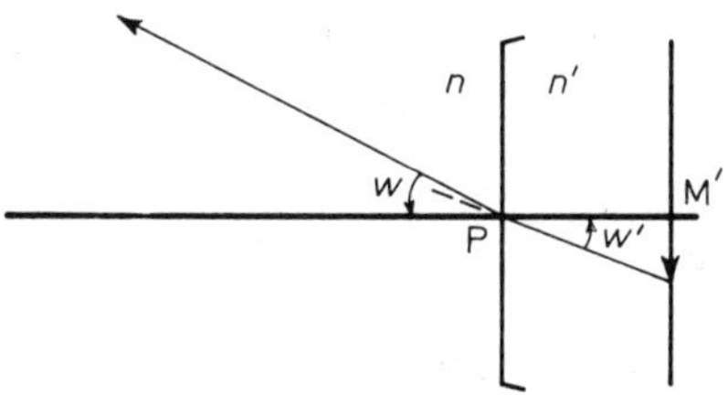


*Figure 8.13 Determination of the apparent size of an object by projection through the centre of the entrance pupil* P

As Pirenne (1967) points out, these approaches to the problem assume that the eye is motionless. Since this is not normally the case, he suggested that 'the best choice of the centre of perspective is the centre of rotation of the eye which is situated near the centre of the eyeball'.

In this geometrical approach object size and apparent size will be identical if object plane and retina are conjugate. If these two planes are not conjugate, the retinal image will be blurred. Assuming that the whole blur circle contributes to the retinal image formation, the total retinal image size will be equal to the basic retinal image plus one blur circle diameter (*see* section 8.3). It has been customary to project this total retinal image back to the object plane (but any other plane could be chosen). If object plane and retina are not conjugate planes, the retinal image when projected back to the object plane will be blurred.

The projection of the basic retinal image through the principal point indicates the position of the centre of the extreme blur circles of the apparent object. The blur circle diameter of the apparent object may be determined on the basis of the pupil diameter, the object distance and the far point distance.

If physical and physiological considerations such as the energy distribution in the retinal blur circle and aspects of contrast are added to the to the geometrical problems set out above, the determination of the apparent size on purely geometrical grounds will become untenable.

A further complication arises from the psychological aspects. The size constancy phenomenon shows that the apparent size of an object does not wholly depend on its retinal image, but also on its apparent distance. The perceived size of an after-image of an object varies with its apparent distance (Kaufman, 1974). Mention must also be made of the apparent size of an object as perceived under conditions of binocular vision (Ogle, 1964).

### 8.5.2 CALCULATIONS

Notwithstanding the problems set out in the previous section, calculations and examination questions have been produced where the geometrical solution has been accepted as giving the correct answer. The following examples deal with such solutions.

#### EXAMPLE 7

Given the details of example 4 (section 8.2), calculate the apparent size of the object, in the object plane.

*Solution 7*

Since the object is a point object, and object plane and retina are not conjugate, the retinal image will be a blur circle. The blur circle now becomes the basis for the determination of the apparent size $j_a$.

From the solution of example 4, the following details are known (*Figure 8.4*):

$$l = -4 \text{ m}$$
$$j = 1.04 \text{ mm}$$
$$k' = +25.63 \text{ mm}$$

Draw a diagram on the basis of the details provided (*Figure 8.14*). The chief rays from the extremities of the retinal image are projected

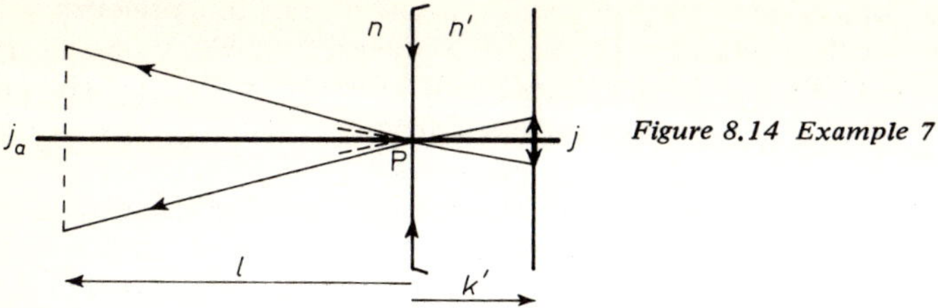


*Figure 8.14 Example 7*

through the principal point of the reduced eye to the object plane whereby the law of refraction is observed. Now,

$$\tan w' = \frac{j}{k'}$$

and, rearranging equation (6.1)

$$\tan w = \frac{4}{3}\tan w'$$

or

$$\frac{j_a}{l} = \frac{4j}{3k'}$$

Hence,

$$j_a = \left|\frac{4jl}{3k'}\right| \tag{8.3}$$

By substitution, the apparent size is

$$\begin{aligned} j_a &= \frac{4 \times 1.04 \times 4000}{3 \times 25.63} \\ &= 216.41 \text{ mm} \end{aligned}$$

Since this measurement is outside the eye, it ought to be expressed in centimetres, in line with normal practice. Thus

$$j_a = 21.64 \text{ cm}$$

Equation (8.3) may also be altered for use with vergences, thus

$$\begin{aligned} j_a &= \frac{4}{3}j \times \frac{1/L}{(4/3)/K'} \\ &= \frac{4/3jK'}{4/3L} \end{aligned}$$

or

$$j_a = \left| j \frac{K'}{L} \right| \qquad (8.4)$$

By substitution (see example 4), the apparent size is calculated to be

$$j_a = 1.04 \times \frac{52.00}{0.25}$$

$$= 216.32 \text{ mm or } 21.63 \text{ cm}$$

The following example will deal with the apparent size of an extended object.

EXAMPLE 8

Given the details of example 6 (section 8.3), calculate the apparent size of the object, in the object plane.

*Solution 8*

The following details were given or calculated earlier:

| | | | | | |
|---|---|---|---|---|---|
| $k'$ | = +25.63 mm | or | $K'$ | = | +52.00 D |
| $l$ | = −4 m | or | $L$ | = | −0.25 D |
| $h'$ | = −0.24 mm | | $j$ | = | 1.04 mm |
| $h$ | = +50 mm | | | | |

As can be seen from *Figure 8.10*, the apparent size of the basic retinal image $h'$ is the same as its object size $h$. Since each point of the apparent image of the basic retinal image is surrounded by an apparent blur of size $j_a$, the total apparent size will be given by

$$|h| + j_a$$

By substitution into either equation (8.3) or (8.4) the apparent blur circle is found to be 210.32 mm. Hence, the total apparent size is 210.32 + 50 = 260.32 mm or 26.03 cm.

NOTE

It is not necessary to memorize equation (8.3), for by using equation (4.3) which is an application of Snell's law, and by inspection of the triangles formed in the diagram (*Figure 8.14*), the problem can be solved.

# Chapter Nine
# The Correction of Spherical Ametropia

## 9.1 General considerations

It has been shown that only an object situated at the far point of an eye will be imaged on that eye's retina (chapter 5). Since the far point of an ametropic eye is not situated at infinity, a distant object will not be imaged sharply on its retina. The question is now: is there an optical arrangement whereby a distant object will be imaged on the retina of an ametropic eye? And the answer: if a distant object can be imaged at the eye's far point, then that image will act as if it were an object situated at the far point of that eye and it will thus be imaged on its retina.

## 9.2 Myopia

The myopic eye's far point is situated in front of the eye (*Figure 5.1*). If there is a distant object it will have to be imaged in the eye's far point,

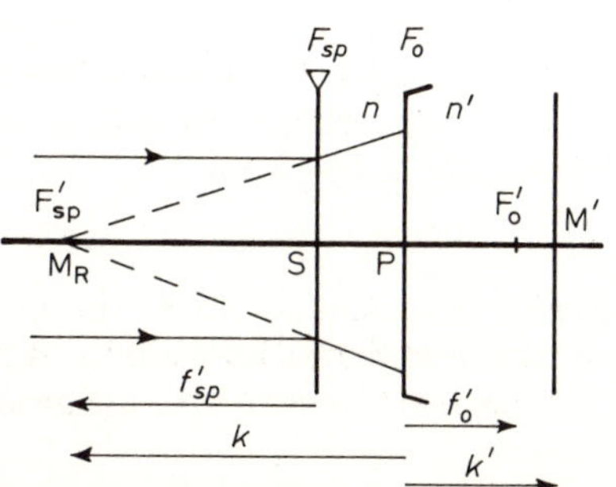


*Figure 9.1 The principles of the correction of ametropia; myopia*

or appear to be imaged there, in order to be 'seen' clearly. This objective may be achieved by means of a negative lens of power $F_{sp}$ (*Figure 9.1*): if the far point of the eye $M_R$ and the second focal point $F'_{sp}$ of that lens coincide, it will appear to the eye when looking through this lens as if the object were situated in its far point plane. Since far point

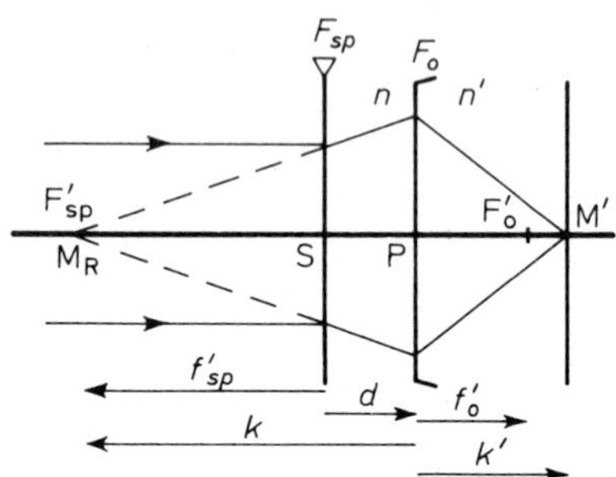


*Figure 9.2 Completion of* Figure 9.1; *the vertex distance d*

plane and retina are conjugate planes, this apparent object will be imaged on the retina (*Figure 9.2*).

### 9.2.1 OCULAR CORRECTION

If an eye has an ocular correction of $K$ dioptres, its far point will be situated $k$ metres away. In order to correct the ametropia a spectacle lens will have to be placed in front of that eye. If the spectacle lens could be placed in touch with the cornea (in case of a reduced eye: could be placed at the reduced surface P), the spectacle correction would have the same power as the ocular correction, that is $F_{sp} = K$. This is (almost) the case when a contact lens correction is fitted on the cornea of a real eye. Accurately, $F_{sp}$ represents the back vertex power of the spectacle lens.

### 9.2.2 SPECTACLE CORRECTION

A spectacle correction cannot normally be fitted on the cornea, but has to be fitted in a spectacle frame, some distance in front of the eye's cornea (in case of a reduced eye, some distance in front of the reduced surface; *Figure 9.2*). The distance between the visual point of a lens (that is, the point of intersection of the visual axis with the back surface

of a lens) and the corneal apex, is known as the vertex distance (BS 3521:1962). This distance should be measured along the axis of the system and is denoted by the letter $d$ (*Figure 9.2*).

NOTE
The vertex distance used in connection with the reduced eye where it is the distance SP (where S denotes the position of the spectacle plane), should not be confused with the clinical measurement of the same name. The latter, in terms of the reduced eye concept, is the distance SA (*Figure 4.3*).

With the sign convention in mind, it can be seen (*Figure 9.2*) that

$$f'_{sp} = k + d \tag{9.1}$$

where $f'_{sp}$ = the second focal length of the spectacle lens (accurately, the back vertex focal length)
$k$ = the far point distance
$d$ = the vertex distance

An eye of given ocular correction may be corrected by any one of a series of lenses having different powers, situated at a vertex distance appropriate for the power of the lens in use. The spectacle lens power may be calculated by either of two methods:

### *9.2.2.1 The 'step-along' method*

This consists of a computing schedule and numerical calculations. The advantage of the method is that it is unnecessary to remember a formula. Mathematical tables, a slide rule or a calculating device will be useful, but are not even necessary. The method follows the following pattern:

1. The vergence of the light incident on the first optical component, and the power of that component, are added to give the emergent vergence. This is an application of the conjugate foci formula.
2. Since the second optical component may be situated some distance away from the first, the effect of the emergent vergence when incident on the second component must be found. This vergence depends on the position of the second component with respect to the focus of the light incident on it. In order to take account of differences in refractive indices of different media, the equivalent air distance of the vergence emerging from the first component is calculated. This is modified by the distance in air between the two optical components.

3. The vergence incident on the second component may now be calculated on the basis of the modified distance in air and the refractive index of the medium between the two optical components.
4. This process may be continued as long as required, using the same schedule.

See also section 1.2, and *Figures 1.16* to *1.20* inclusive.

The method may be set out as follows:

$$\begin{array}{llll} L_1 & & & \\ F_1 & + & & \\ \hline L_1' & & \longrightarrow & l_1' \\ & & & d \quad - \\ \cline{4-4} L_2 & & \longleftarrow & l_2 \\ F_2 & + & & \\ \cline{1-1} L_2' & & & \end{array}$$

NOTE

The vertex distance is always a positive quantity because it is measured from the first to the second lens, that is, from left to right.

In view of the reversibility of the ray path, this method may also be used in the opposite direction. Vergences which originally emerged from a lens now become vergences incident on a lens; for instance,

$$\begin{array}{ll} L_1' & \\ F_1 & + \\ \hline L_1 & \end{array}$$

This latter approach may be used to calculate the spectacle correction from the ocular correction and the vertex distance. The ocular correction may then be likened to the vergence emerging from the eye, and the schedule is as follows (*Figure 9.2*):

$$\begin{array}{lcl} K & \longrightarrow & k \\ & & d \quad + \\ \cline{3-3} F_{sp} & \longleftarrow & f_{sp}' \end{array}$$

### 9.2.2.2 *The effectivity formula*

The 'step-along' method may be written as a formula derived as follows. Consider *Figure 9.3*, first with respect to the initial lens of power $F_1$.

The focal length will be $f'_1$ metres (assuming $n = 1$). If the incident light is parallel, the image will be formed at $F'_1$, the second principal focus.

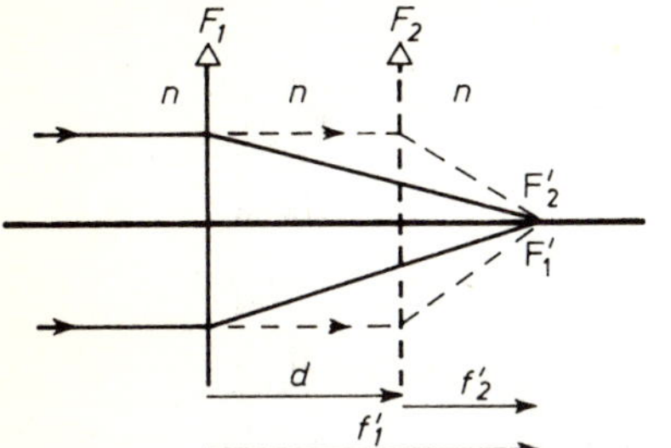


*Figure 9.3 The effectivity formula*

Now consider the system with respect to the lens of power $F_2$. It is assumed that the first lens of power $F_1$ has been removed and that the image will still be formed in the same plane, that is at $F'_1$. Therefore, the foci $F'_1$ and $F'_2$ must coincide. The second focal length of lens number 2 will be equal to $f'_2$ metres.

The relationship between $f'_2$ and $f'_1$ is given by

$$f'_2 = f'_1 - d$$

where $d$ is the distance between the two lenses.

Thus

$$F_2 = \frac{1}{f'_2}$$

or, by substitution,

$$F_2 = \frac{1}{f'_1 - d} \tag{9.2}$$

The application of this general formula in visual optics is the following. $F_1$ may be substituted by $F_{sp}$, and $F_2$ by $K$. Equation (9.2) will now read

$$K = \frac{1}{f'_{sp} - d} \tag{9.3}$$

where $d$ is the vertex distance.

Since the spectacle correction is commonly known in dioptres, this

equation may be modified so that the ocular correction is also found in dioptres, thus

$$K = \frac{1}{(1/F_{sp}) - d}$$

$$= \frac{1}{(1 - dF_{sp})/F_{sp}}$$

$$= \frac{F_{sp}}{1 - dF_{sp}} \qquad (9.4)$$

The reader should not find it too difficult to derive in a similar manner the equation for the spectacle correction when the vertex distance and the ocular correction are known. The equation is

$$F_{sp} = \frac{1}{k + d} \qquad (9.5)$$

and hence

$$F_{sp} = \frac{K}{1 + dK} \qquad (9.6)$$

### *9.2.2.3 Calculations*

EXAMPLE 9

A reduced eye has an ocular correction of −10.00 D. Calculate the spectacle correction for a vertex distance of 10 mm.

*Solution 9*

*Step 1.* Draw a diagram on the basis of the data provided (*Figure 9.4*).

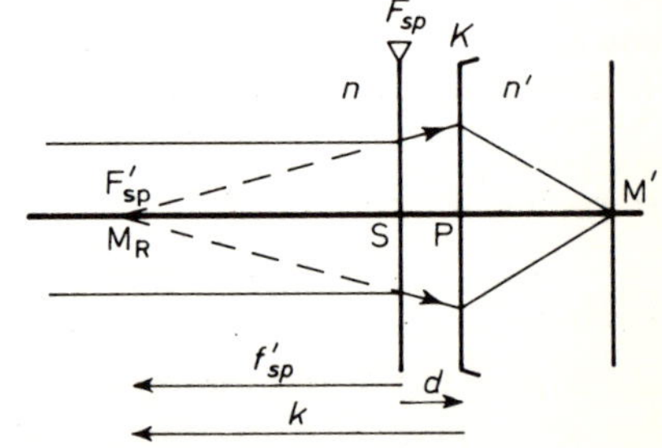


***Figure 9.4** Examples 9 and 11*

*Step 2.* Calculate the spectacle correction:

(a) Using the 'step-along' method.

$$K = -10.00\text{ D} \longrightarrow k = -10.00\text{ cm}$$

$$d = +1.0 \quad +$$

$$F_{sp} = -11.11 \longleftarrow f'_{sp} = -9.00$$

NOTE
The vertex distance is normally given in millimetres. Since the calculation deals with distance measured outside the eye, one must not forget to convert the vertex distance from millimetres to centimetres.

(b) Using the effectivity formula (equation (9.6)).

$$F_{sp} = \frac{-10.00}{1 + 0.01(-10.00)}$$

$$= \frac{-10.00}{1 - 0.10}$$

$$= -11.11 \text{ D}$$

NOTE
Since this formula is expressed in dioptres, the vertex distance must be expressed in metres.

NOTE
The far point is conjugate with infinity through the spectacle correction.

EXAMPLE 10
Calculate the ocular correction of an eye which is corrected by a spectacle lens of −6.50 D at 13 mm.

*Solution 10*
*Step 1.* Draw a diagram (*see Figure 9.4*).
*Step 2.* Calculate the ocular correction:

(a) Using the 'step-along' method

$$\begin{array}{lllll} F_{sp} = -6.50 \text{ D} & \longrightarrow & f'_{sp} & = & -15.38 \text{ cm} \\ & & d & = & +1.3 \quad - \\ \hline K = -6.00 & \longleftarrow & k & = & -16.68 \end{array}$$

(b) Using the effectivity formula (equation (9.4))

$$K = \frac{-6.50}{1 - 0.013\,(-6.50)}$$

$$= \frac{-6.50}{1 + 0.08}$$

$$= -6.02 \text{ D}$$

*9.2.2.4 The vertex distance equation*

When both the ocular correction and the spectacle correction are known, the vertex distance may be calculated from equation (9.1). This, however, requires that the far point distance and the second focal length of the spectacle lens are calculated first.

Alternatively, the vertex distance may be calculated from an equation which may be derived as follows (*Figure 9.4*):

From equation (9.4):

$$K = \frac{F_{sp}}{1 - dF_{sp}}$$

or

$$1 - dF_{sp} = \frac{F_{sp}}{K}$$

Now

$$-dF_{sp} = \frac{F_{sp}}{K} - 1$$

$$= \frac{F_{sp} - K}{K}$$

Hence

$$d = \frac{K - F_{sp}}{F_{sp}K} \qquad (9.7)$$

From equation (9.6):

$$F_{sp} = \frac{K}{1 + dK}$$

or

$$1 + dK = \frac{K}{F_{sp}}$$

Now

$$+dK = \frac{K}{F_{sp}} - 1$$

$$= \frac{K - F_{sp}}{F_{sp}}$$

Hence

$$d = \frac{K - F_{sp}}{F_{sp}K} \qquad (9.7)$$

EXAMPLE 11

An ametropic eye has an ocular correction of −10.00 D and is corrected by a spectacle lens of power −11.75 D. Calculate the vertex distance required to correct this eye for distance vision.

*Solution 11*

*Step 1.* Draw a diagram (*Figure 9.4*).

*Step 2.* The vertex distance may be found from either equation (9.1) or (9.7).

Using equation (9.1):

$$F_{sp} = -11.75\text{ D} \longrightarrow f'_{sp} = -8.51\text{ cm}$$
$$K = -10.00 \longrightarrow k = -10.00 \quad -$$
$$d = +\ 1.49\text{ cm} = +14.9\text{ mm}$$

Using equation (9.7):

$$d = \frac{-10.00-(-11.75)}{-11.75\ (-10.00)}$$
$$= \frac{+1.75}{+117.50}$$
$$= +0.014\,9\text{ m} = +14.9\text{ mm}$$

NOTE

Since the unit used in the equation is the dioptre, the vertex distance will be calculated in metres.

In conclusion, an ametropic eye of −10.00 D ocular correction may be corrected by a spectacle lens of −11.75 D at a vertex distance of 14.9 mm as well as by a spectacle lens of −11.11 D at 10 mm, or any other suitable combination. However, the size of the retinal image is affected by varying amounts depending on the combination used. This aspect will be discussed in chapter 10.

## 9.3 Hyperopia

The hyperopic eye's far point is situated behind the eye (*Figure 6.4*). For a distant object to be 'seen' clearly it will have to be imaged in the

far point plane of the eye. In turn, this far point plane image will act as object for the eye. Since far point plane and retina are conjugate planes, the image will be clearly focussed on the retina.

### 9.3.1 SPECTACLE CORRECTION

This above stated objective will be achieved when a positive spectacle lens of power $F_{sp}$ is placed in front of the eye such that its second focal length $f'_{sp}$ is equal to the distance from the spectacle plane to the far point plane of the eye. The second principal focus $F'_{sp}$ of the spectacle lens and the far point $M_R$ are thus coincident. Now, a distant object will be imaged in the far point plane of the onlooking eye and, in turn, it will be imaged on the retina of that eye.

Consider the following example:

**EXAMPLE 12**

A reduced eye has an ocular correction of +10.00 D. Calculate the power of the spectacle lens which will correct the eye for distance vision when placed at a vertex distance of 10 mm.

*Solution 12*

*Step 1.* Draw a diagram (*Figure 9.5*).

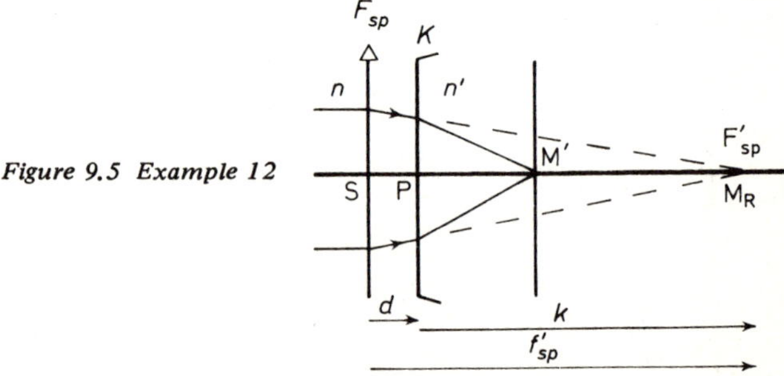


*Figure 9.5 Example 12*

*Step 2.* Calculate the power of the spectacle lens required.
Using the 'step-along' method:

$$K = +10.00\text{ D} \longrightarrow k = +10.00\text{ cm}$$
$$d = +1.00 \quad +$$
$$F_{sp} = +9.09 \longleftarrow f'_{sp} = +11.00$$

Using the effectivity formula, equation (9.6);

$$F_{sp} = \frac{K}{1 + dK}$$

$$= \frac{+10.00}{1 + 0.01\ (+10.00)}$$

$$= \frac{+10.00}{1 + 0.1}$$

$$= \frac{+10.00}{+1.1}$$

$$= +9.09\ \text{D}$$

Like the myopic eye, the hyperopic eye may be corrected by any one of a number of lens powers provided the lens is placed at the appropriate vertex distance. As an example consider the following problem:

EXAMPLE 13

A reduced eye has an ocular correction of +10.00 D. At which vertex distance should a spectacle lens of +8.50 D be placed so that it will correct this eye for distance vision?

*Solution 13*

*Step 1.* Draw a diagram (*Figure 9.5*).

*Step 2.* Calculate the vertex distance:

(a) Using the 'step-along' method (equation (9.1)):

$$F_{sp} = +8.50\ \text{D} \longrightarrow f'_{sp} = +11.76\ \text{cm}$$
$$K = +10.00 \longrightarrow k = +10.00\ \ -$$
$$d = +1.76\ \text{cm} = +17.6\ \text{mm}$$

(b) Using the vertex distance equation, equation (9.7):

$$d = \frac{K - F_{sp}}{F_{sp}K}$$

$$= \frac{+10.00 - (+8.50)}{+8.50\ (+10.00)}$$

$$= \frac{+1.50}{+85.00}$$

$$= +0.017\,6\ \text{m} = +17.6\ \text{mm}$$

## 9.4 Effect of a vertex distance change

It is true to say that the effective power of a spectacle correction at the eye (or at the reduced surface of a reduced eye) must be equal to the ocular correction. If the spectacle plane is moved from $S_1$ to $S_2$

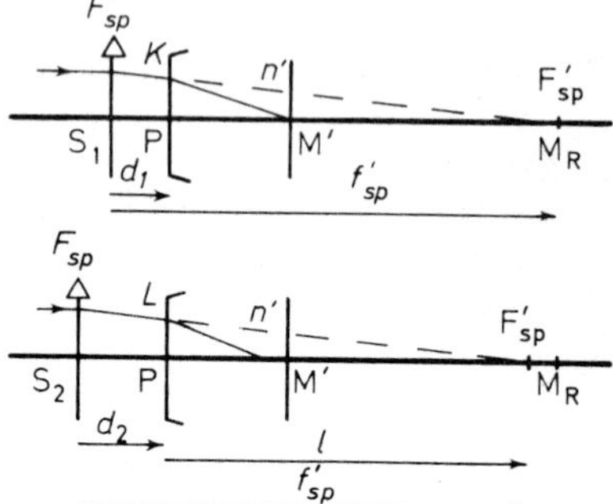


*Figure 9.6 Example 14; the effect of a change in vertex distance: positive correction*

(*Figure 9.6*), the difference in power of the spectacle correction required to fulfil this condition is denoted $\delta F_{sp}$, and is given by

$$\delta F_{sp} = F_{sp2} - F_{sp1} \tag{9.8}$$

where $F_{sp1}$ = the spectacle correction in position 1
$F_{sp2}$ = the spectacle correction in position 2

With regard to equation (9.6), the spectacle correction in position 2 may be given in terms of the spectacle correction in position 1, thus

$$F_{sp2} = \frac{F_{sp1}}{1 + \delta d F_{sp1}}$$

where $\delta d$ = the distance between the two spectacle points, that is $d_2 - d_1$. Hence,

$$\begin{aligned}
\delta F_{sp} &= \frac{F_{sp1}}{1 + \delta d F_{sp1}} - F_{sp1} \\
&= \frac{F_{sp1} - F_{sp1}(1 + \delta d F_{sp1})}{1 + \delta d F_{sp1}} \\
&= \frac{F_{sp1} - F_{sp1} - \delta d F_{sp1}^2}{1 + \delta d F_{sp1}} \\
&= \frac{-\delta d F_{sp1}^2}{1 + \delta d F_{sp1}}
\end{aligned}$$

Since $\delta d$ must be entered in metres, it will be a very small fraction indeed. Hence, the factor $(\delta d F_{sp1})$ may be ignored and so the denominator becomes unity. Consequently, the approximate change in spectacle correction required is given by

$$\delta F_{sp} = -\delta d F_{sp1}^2 \qquad (9.9)$$

This equation may be tested on the next two examples.

EXAMPLE 14

From example 12, $F_{sp1}$ = +9.09 D at $d_1$ = +10 mm, and
from example 13, $F_{sp2}$ = +8.50 D at $d_2$ = +17.6 mm.
Using equation (9.8),

$$\begin{aligned} \delta F_{sp} &= +8.50 - (+9.09) \\ &= -0.59 \text{ D} \end{aligned}$$

Using the approximate formula, equation (9.9),

$$\begin{aligned} \delta F_{sp} &= -(0.0176 - 0.01)(+9.09)^2 \\ &= -(0.0076)(+82.63) \\ &= -0.63 \text{ D} \end{aligned}$$

EXAMPLE 15

From example 9, $F_{sp1}$ = −11.11 D at $d_1$ = +10 mm. Similarly,
from example 11, $F_{sp2}$ = −11.75 D at $d_2$ = +14.9 mm.
Using equation (9.8),

$$\begin{aligned} \delta F_{sp} &= -11.75 - (-11.11) \\ &= -0.64 \text{ D} \end{aligned}$$

Using the approximate formula, equation (9.9),

$$\begin{aligned} \delta F_{sp} &= -(0.0149 - 0.01)(-11.75)^2 \\ &= -(0.0049)(+139.24) \\ &= -0.68 \text{ D} \end{aligned}$$

NOTE

When a vertex distance is increased, the spectacle correction must be decreased (i.e. be made more negative in power). The opposite is true when the vertex distance is reduced.

However, in practice these changes are not always made. This raises the question: what is the effect at the eye of increasing the vertex distance without an appropriate change in the spectacle correction? Take as an example the situation of example 15 (*Figure 9.7*). When the

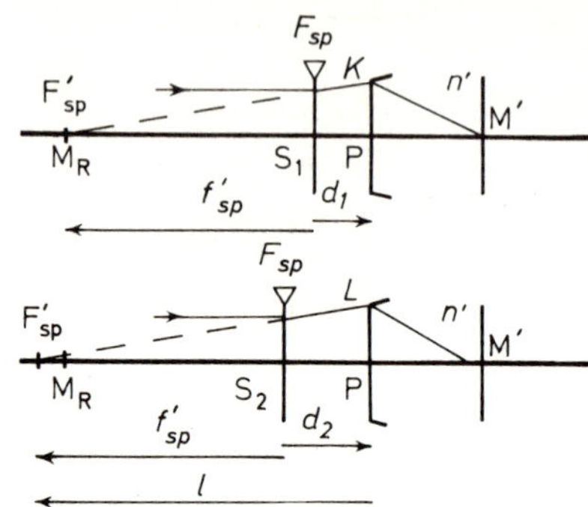


*Figure 9.7 Example 15; the effect of a change in vertex distance: negative correction*

original spectacle correction $F_{sp1} = -11.11$ D is moved from $d_1 = +10$ mm to $d_2 = +14.9$ mm, the vergence incident on the eye changes. When the spectacle correction was at the original spectacle point, the vergence at the eye was equal to the ocular correction, that is $K = -10.00$ D. When placed at $S_2$, the vertex distance is $d_2$ and the vergence incident on the eye, $L$, is:

$$F_{sp1} = -11.11\text{ D} \longrightarrow f'_{sp1} = -9.00\text{ cm}$$
$$d_2 = +1.49 \quad -$$
$$L = -9.53 \longleftarrow l = -10.49$$

However, it should have been equal to $K$. Hence, the eye is now under-corrected, namely by

$$L - K$$
$$= -9.53 - (-10.00)$$
$$= +0.47\text{ D}$$

As for the situation of example 14, when $F_{sp1} = +9.09$ D is moved

from the original spectacle point to $d_2$ = +17.6 mm, the vergence incident on the eye becomes (*Figure 9.6*)

$$
\begin{array}{lcllcl}
F_{sp1} & = & +9.09 \text{ D} & \longrightarrow f'_{sp1} & = & +11.00 \text{ cm} \\
 & & & \quad d_2 & = & +1.76 \quad - \\
\hline
L & = & +10.82 & \longleftarrow \; l & = & +9.24
\end{array}
$$

This eye has now become overcorrected, namely by

$$
\begin{aligned}
& L - K \\
= \; & +10.82 - (+10.00) \\
= \; & +0.82 \text{ D}
\end{aligned}
$$

NOTE

When the vertex distance of a given spectacle correction is increased (and the power of the correction is not adjusted), the effectivity of the correction at the eye increases too (becomes more positive). This means that the eye will then be able to see clearly a point, nearer than the far point. This has implications for near vision and presbyopia (chapter 13).

# Chapter Ten
# Spectacle Magnification

## 10.1 Definition

Spectacle magnification may be defined as the ratio of the retinal image size of an object in the corrected ametropic eye to the blurred or sharp retinal image size of that object in the ametropic eye when uncorrected.

NOTE
The words 'blurred or sharp' appear in the classical form of the definition. The blurred dimensions of the retinal image will depend on the pupil diameter. Since the pupil diameter is a variable quantity it is not normally taken into account. Hence, a practical definition of the spectacle magnification is

$$SM = \frac{\text{the retinal image size of an object in the corrected ametropic eye}}{\text{the basic retinal image size of that object in the ametropic eye}}$$

Thus, if these retinal image sizes can be calculated, for instance, from first principles, it would not really be necessary to derive a formula. When the formulae have slipped one's mind, one could resort to this approach.

## 10.2 Derivation of formulae

NOTE
The formulae to be derived below are only part of the spectacle magnification. All formulae provide only the so-called power factor. This is the magnification due to the focal power of the lens. The associated factor, known as the shape factor, will be discussed in section 14.4. Since the shape factor is very small for the most frequently used spectacle lenses, it will be ignored for the moment.

The formula for the spectacle magnification may be derived as follows. Consider an uncorrected myopic eye (*Figure 10.1*). A ray originating from the top of a distant object is incident as chief ray under an angle

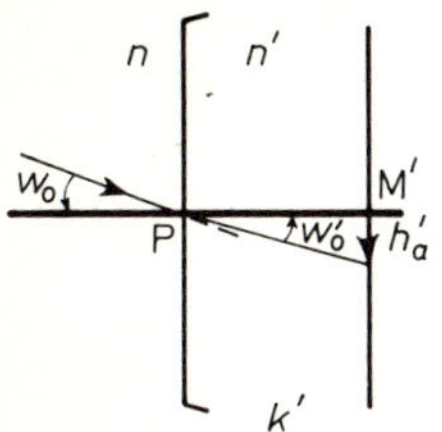


*Figure 10.1 The basic retinal image in an uncorrected myopic eye*

$w_o$ at the reduced surface P. The basic retinal image size $h_a'$ in this ametropic eye is then given by (equation (4.3))

$$h_a' = -k' \frac{3}{4} \tan w_o \qquad (10.1)$$

When this eye is corrected by a spectacle lens of power $F_{sp}$ at a vertex distance $d$, the angle subtended by the chief ray from the top of that distant object, at the spectacle plane S, is also $w_o$ (*Figure 10.2*).

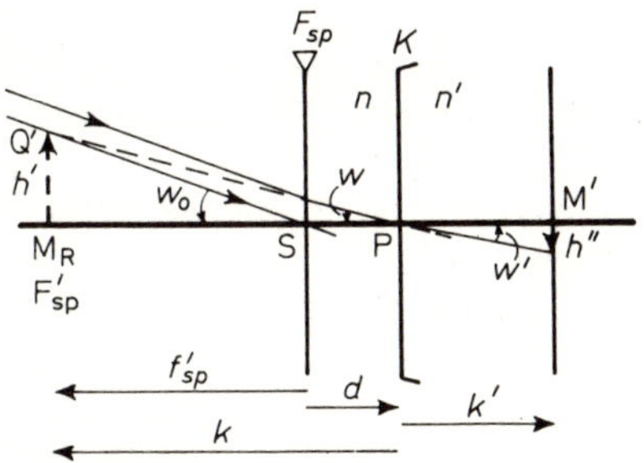


*Figure 10.2 Spectacle magnification: the retinal image formation of a distant object in a corrected myopic eye*

By definition, the spectacle lens forms an image $h'$ of the distant object in the far point plane $M_R$. All rays incident on the reduced surface of the eye appear to originate from the far point plane image, which now acts as object for the eye. Consequently, the chief ray incident on the reduced surface may be drawn (*Figure 10.2*). This is the ray Q′P, which is incident under an angle $w$ at the reduced surface. Now

$$\tan w = \frac{-h'}{k} \tag{10.2}$$

and

$$\tan w_o = \frac{-h'}{f'_{sp}} \tag{10.3}$$

The retinal image size $h''$ in this corrected ametropic eye may be calculated from

$$\tan w' = \frac{-h''}{k'}$$

or

$$h'' = -k' \tan w'$$

$$= -k' \frac{3}{4} \tan w \tag{10.4}$$

According to the definition, spectacle magnification is given by

$$SM = \frac{h''}{h'_a}$$

Substituting equations (10.4) and (10.1) into this formula,

$$SM = \frac{k' \frac{3}{4} \tan w}{k' \frac{3}{4} \tan w_o}$$

$$= \frac{\tan w}{\tan w_o}$$

From equations (10.2) and (10.3),

$$SM = \frac{-h'}{k} \Big/ \frac{-h'}{f'_{sp}}$$

$$= \frac{h' f'_{sp}}{k h'}$$

$$= \frac{f'_{sp}}{k} \tag{10.5}$$

or

$$SM = \frac{K}{F_{sp}} \tag{10.6}$$

It might be easier to remember the following formula which is based on equation (10.5)

$$SM = \frac{\mathrm{SM_R}}{\mathrm{PM_R}} \tag{10.7}$$

*See Figure 10.2.*

NOTE

Since the effect of the ametropic eye on the basic retinal image size is the same whether the eye be corrected or not, the ratio of the retinal image sizes is affected only by the difference in the angle subtended by the chief rays at the reduced surface. Hence, it is an angular magnification.

In view of the above Note, spectacle magnification may also be defined as (*Figure 10.2*)

$$SM = \frac{w}{w_o}$$

Substituting equations (10.2) and (10.3), equation (10.5) is found.

Other formulae for the spectacle magnification may be derived as follows. From equations (10.6) and (9.7)

$$\frac{1}{SM} = 1 - dF_{sp}$$

or

$$SM = \frac{1}{1 - dF_{sp}} \tag{10.8}$$

When this division is developed, a binomial formula is found which may be simplified to the approximate formula of

$$SM = 1 + dF_{sp} \tag{10.9}$$

(neglecting higher powers of $dF_{sp}$).

Hence, the approximated percentage change produced by the spectacle correction in the retinal image size, is given by

$$SM_{\%} = dF_{sp} \tag{10.10}$$

where $d$ is entered in centimetres.

From (10.6) and (9.7) a formula which resembles equation (10.9) is obtained. This formula, however, gives the accurate spectacle magnification:

$$SM = 1 + dK \qquad (10.11)$$

## 10.3 Calculations

NOTE
In order to find by calculation very nearly the same image size, using different methods, it is necessary to calculate the (spectacle) magnification to four decimal places, as will be seen below.

Using equation (10.5), the spectacle magnification in example 11 ($f'_{sp} = -8.51$ cm; $k = -10.00$ cm) is found to be

$$SM = \frac{-8.51}{-10.00} = +0.851\,0$$

Similarly for example 12 ($f'_{sp} = +11.00$ cm; $k = +10.00$ cm; $d = +10$ mm)

$$SM = \frac{+11.00}{+10.00} = +1.100\,0$$

and for example 13 ($f'_{sp} = +11.76$ cm; $k = +10.00$ cm; $d = +17.6$ mm)

$$SM = \frac{+11.76}{+10.00} = +1.176\,0$$

NOTE
From these calculations it can be seen that spectacle magnification is always a positive value; that in myopia its value is smaller than unity and in hyperopia larger than unity. This means that the corrected myopic eye has a smaller retinal image and the corrected hyperopic eye a larger retinal image, than each eye's basic retinal image.

NOTE
It will be seen from these examples that for the same ocular correction the larger the vertex distance the larger the departure of the spectacle magnification from unit value. A summary is given in section 11.5.

Graves (1968) gives as rule of thumb that for a vertex distance of 14 mm the spectacle magnification is 1.4% per dioptre of spectacle lens

power. Note that low corrections (or small changes) give small *SM*-factors (or small changes in spectacle magnification).

## 10.4 Practical importance

The importance of the spectacle magnification is situated in the comparison of retinal image sizes before and after correction, or after a change in the correction. A newly corrected hyperope will find familiar objects to appear larger than they used to be. Since a 'bigger' object is associated with 'being nearer' (one of the clues used in the monocular perception of relative depth), the ability to judge distances will be somewhat upset. However, this does not normally last for many days.

The newly corrected myope experiences a different change. Since the retinal image of the corrected myopic eye is smaller than that of the uncorrected eye, objects appear to be further away than they used to be. This could turn out to be fatal when driving a car! Patients should be made aware of this problem. Usually, the brain will adapt to the new situation in a matter of days.

## 10.5 Calculation of the retinal image size

The retinal image size of a distant object may be calculated in a number of ways. The solutions will be shown on the basis of the following example.

EXAMPLE 16

A distant object subtends an angle $w_o$ at an axially ametropic reduced eye of ocular correction +5.00 D. This eye is corrected by a spectacle lens at 15 mm. Calculate the retinal image size.

*Solution 16*

### 10.5.1 USING FIRST PRINCIPLES: GEOMETRY

*Step 1.* Draw a diagram based on the information provided (*Figure 10.3*).

NOTE

Since this is a distant object, the angle it subtends at the eye is the same as it subtends at the spectacle point. *See Figure 10.3.*

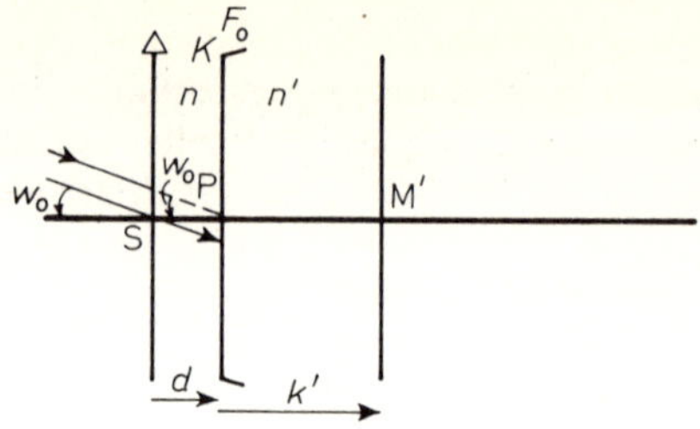


*Figure 10.3 Example 16*

*Step 2.* Calculate the spectacle correction. Using, for instance, the 'step-along' method:

$$K = +5.00\text{ D} \longrightarrow k = +20.00\text{ cm}$$
$$d = +1.5 \quad +$$
$$F_{sp} = +4.65 \longleftarrow f'_{sp} = +21.50$$

*Step 3.* Indicate on the diagram (*Figure 10.4*) the position of the far

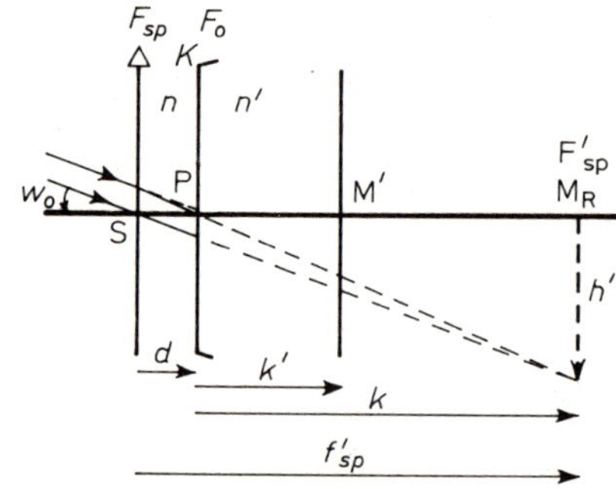


*Figure 10.4 Example 16; formation of the far point plane image of a corrected hyperopic eye*

point $M_R$ and the size of the far point image $h'$. This measures (equation (1.16)):

$$h' = -21.50 \tan w_o \text{ cm}$$

*Step 4.* The far point image now acts as if it were an object with respect to the next optical component, that is, the reduced surface of the eye.

NOTE
Since far point plane and retina are conjugate planes, the far point image will be focussed on the retina. Hence, no further optical calculations to find the image position with respect to the retina are necessary.

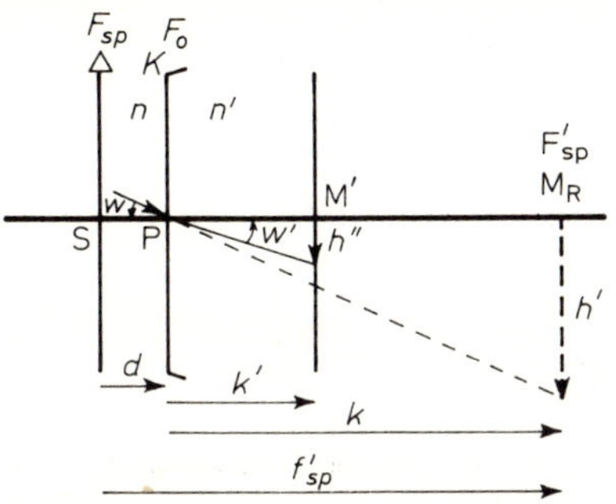


*Figure 10.5 Example 16; the formation of the retinal image*

The far point image $h'$ subtends an angle $w$ at the eye where (*Figure 10*

$$\tan w = \frac{-h'}{k}$$

*Step 5.* The retinal image $h''$ may be calculated as follows:
The angle $w'$ subtended by the retinal image at the reduced surface, is

$$\tan w' = \frac{-h''}{k'}$$

or

$$h'' = -k' \tan w'$$

By successive substitution, the retinal image size is found to be

$$h'' = -k'\frac{3}{4}\tan w$$

$$= -k'\left(\frac{3}{4}\right)\left(\frac{-h'}{k}\right)$$

$$= \frac{3k'h'}{4k}$$

*Step 6.* Calculate the axial length of this eye. The dioptric length may be found from the ametropia formula (equation (5.5)):

$$K' = +5.00 + 60.00$$

$$= +65.00 \text{ D}$$

and the axial length is then (equation (5.1))

$$k' = +20.51 \text{ mm}$$

*Step 7.* By substitution the retinal image size is found to be (from Step 5)

$$h'' = \frac{3 \times 20.51 \times (-215.00 \tan w_o)}{4 \times 200.00}$$

$$= -16.54 \tan w_o \text{ mm}$$

### 10.5.2 USING FIRST PRINCIPLES: MAGNIFICATION

Example 16 may also be solved using the magnification equation (equation (1.15)). However, since the vergence incident on the spectacle lens is zero, this equation cannot be used for the magnification due to the spectacle lens. Hence, up to and including Step 3 (section 10.5.1), the solution is the same. Thence:

*Step 4.* Calculate the magnification due to the optical system of the eye, that is, employing equation (6.3). Substituting the value for the dioptric length of the eye from Step 6 (section 10.5.1),

$$m = \frac{+5.00}{+65.00}$$

$$= +0.076\,9$$

*Step 5.* The retinal image size is given by

$$h'' = mh'$$

$$= +0.076\,9(-215.00 \tan w_o)$$

$$= -16.53 \tan w_o \text{ mm}$$

### 10.5.3 USING SPECTACLE MAGNIFICATION

Since the object is a long distance away, example 16 may also be solved by means of the spectacle magnification. Since the latter must be applied in conjunction with the basic retinal image size, this must first be found.

*Step 1. See Figure 7.2.* The basic retinal image size may be calculated

from equation (4.3). Substituting the axial length (from Step 6, section 10.5.1)

$$h' = -20.51 \times \frac{3}{4} \tan w_o$$

$$= -15.38 \tan w_o \text{ mm}$$

*Step 2.* Calculate the spectacle magnification. Substituting the values from Step 2 (section 10.5.1) into, for instance, equation (10.5),

$$SM = \frac{+21.50}{+20.00}$$

$$= +1.075\,0$$

*Step 3.* The retinal image size is

$$h'' = SM \times h'$$

$$= +1.075\,0(-15.38 \tan w_o)$$

$$= -16.53 \tan w_o \text{ mm}$$

### 10.5.4 USING THE EQUIVALENT POWER

If the corrected ametropic eye is considered as an optical system consisting of two components separated by air, the following approach may be adopted.

By definition, the image of the distant object is formed on the retina. The equivalent power of two (thin) lenses in air, is given by equation (1.18). Substituting $F_{sp}$ for the first lens and $F_o$ for the second lens, the equivalent power of an axially ametropic eye is given by

$$F = F_{sp} + F_o - dF_{sp}F_o$$

If the ametropic eye were subject to refractive ametropia, the formula will read

$$F = F_{sp} + F_e - dF_{sp}F_e$$

$h''$. Since the retina of the eye coincides with the equivalent second focus of the system, equation (4.3) may now be written as

$$h'' = -f'\frac{3}{4}\tan w_o \qquad (10.12)$$

or

$$h'' = -\frac{\tan w_o}{F} \qquad (10.13)$$

Substituting the values applying to example 16 into this expression

$$h'' = -\frac{\tan w_0}{+4.65 + 60.00 - 0.015(+4.65)(+60.00)}$$

$$= -\frac{\tan w_o \times 1000}{+60.46}\ \text{mm}$$

$$= -16.54 \tan w_o \ \text{mm}$$

This approach employs the same ideas as embodied in Jalie's (1977) concept of the 'equivalent surface' discussed in section 4.2. Referring to *Figure 4.1*, equation (10.13) can be shown to apply, as follows.

The distant object (example 16) subtends an angle $w_o$ at both F and N′ (*Figure 4.1*). Hence,

$$\tan w_o = \frac{-\text{F}'\text{Q}'}{\text{N}'\text{F}'}$$

$$= \frac{-\text{F}'\text{Q}'}{-f}$$

which can be substituted into equation (10.13), thus

$$h'' = -\left(\frac{-\text{F}'\text{Q}'}{-f}\right)\Big/F$$

$$= -\left(\frac{-\text{F}'\text{Q}'}{-f}\right)\Big/\frac{-n}{f}$$

$$= -\left(\frac{-\text{F}'\text{Q}'}{-f}\right)\left(\frac{f}{-n}\right)$$

Since $n = 1$,

$$h'' = \text{F}'\text{Q}'$$

## 10.6 Nomogram

*Figure 10.6* is a nomogram which gives the spectacle magnification of any combination of spectacle lens correction between ±15.00 D and

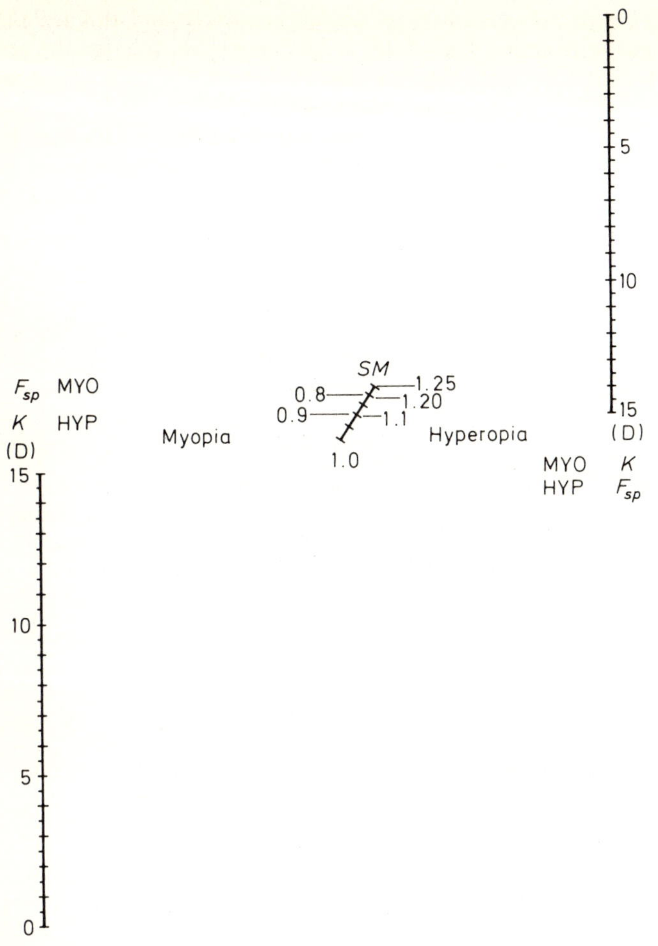


*Figure 10.6 Nomogram providing the spectacle magnification-value for corrections from 0 to + and −15 D*

ocular correction between ±15.00 D. The nomogram may also be used to find the spectacle lens correction which will give a desired spectacle magnification for a given ocular correction. The vertex distance of this correction will have to be calculated separately. For this purpose equation (9.1) or (9.10) can be used.

The nomogram may be used as follows. Assume the spectacle lens has a power of −10.00 D and the ocular correction is −9.00 D. Place a

straight-edge at $F_{sp}$ is 10.00 on the left hand scale and at $K$ is 9.00 on the right hand scale. The straight-edge is now seen to cross the centre scale at 0.9, 1.1. Since this is a case of myopia, the value of the myopia-side of the scale is the value sought. Hence, $SM = +0.9$.

*Figure 10.7* is the same nomogram, but the section with values from 10.00 to 15.00, for aphakia and high myopia, has been redrawn on a larger scale. This increases the accuracy which may be achieved. Note

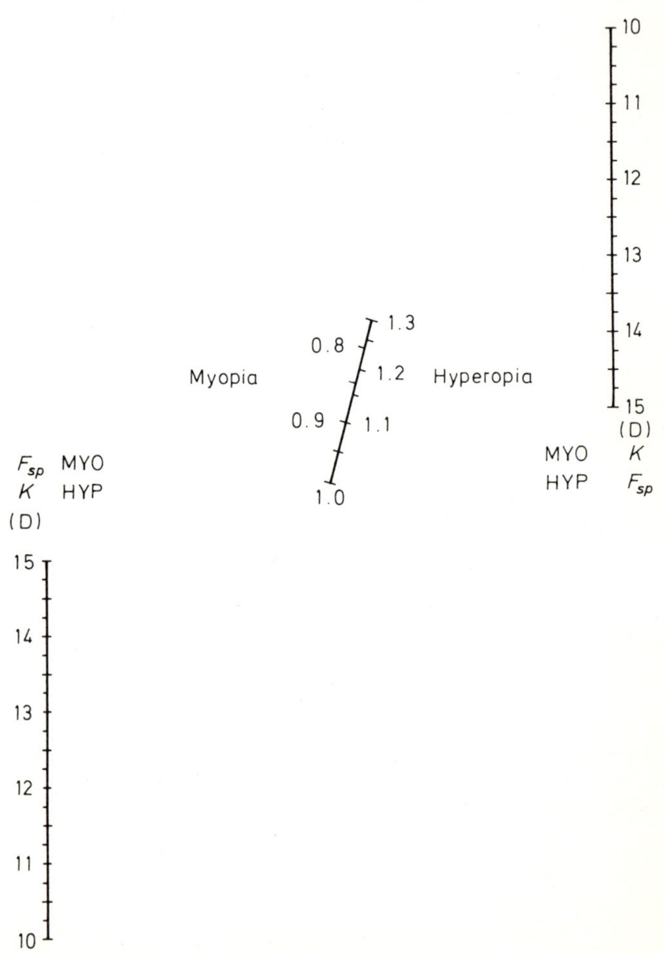


*Figure 10.7 As* Figure 10.6, *for corrections from +10 to +15 and from −10 to −15 D*

that the centre scale differs from the one in *Figure 10.6*. (*See* Obstfeld, 1975).

## 10.7 Universal formula for spectacle magnification

In section 10.2, spectacle magnification was shown to be a case of angular magnification. Thus, in an uncorrected ametropic eye an object

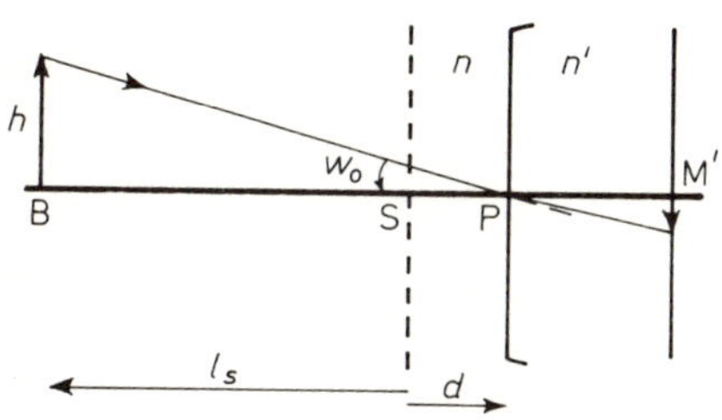


*Figure 10.8 The angle $w_o$ subtended by an object at the uncorrected eye*

situated at some distance from that eye will subtend an angle $w_o$ at the eye (*Figure 10.8*). The angle $w_o$ (in radians) is given by

$$w_o = \frac{h}{l_s - d}$$

$$= \frac{h/(1 - dL_s)}{L_s}$$

$$= \frac{hL_s}{1 - dL_s}$$

When the ametropic eye is corrected, the object subtends an angle $w_s$ (*Figure 10.9*) at the spectacle point S. Assuming this to be a case of myopia, the object at B will be imaged at B′. This image, in turn, will act as object for the eye where it subtends an angle $w$. The angle $w$ is given by

$$w = \frac{-h'}{l}$$

$$= \frac{-h'}{l_s' - d}$$

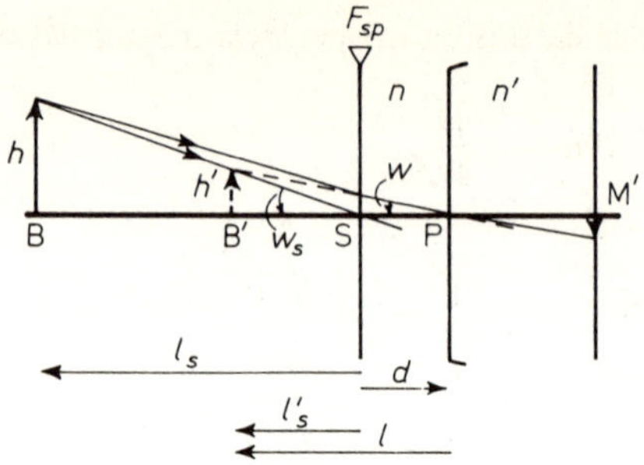

*Figure 10.9 The angle $w_s$ subtended by an object at the spectacle point*

From the magnification equation (equation (1.15)) the size $h'$ is given by

$$h' = mh$$

$$= \frac{L_s}{L'_s}h$$

Hence,

$$w = \frac{L_s}{L'_s}h/(l'_s - d)$$

$$= \frac{L_s h}{L'_s} \times \frac{1}{l'_s - d}$$

$$= \frac{L_s h}{L'_s} \times \frac{1}{\dfrac{1 - dL'_s}{L'_s}}$$

$$= \frac{L_s h}{L'_s} \times \frac{L'_s}{1 - dL'_s}$$

$$= \frac{L_s h}{1 - dL'_s}$$

From the 'angular' definition of spectacle magnification,

$$SM = \frac{w}{w_o}$$

$$= \frac{L_s h}{1 - dL_s'} \bigg/ \frac{hL_s}{1 - dL_s}$$

$$= \frac{L_s h(1 - dL_s)}{hL_s(1 - dL_s')}$$

$$= \frac{1 - dL_s}{1 - dL_s'} \qquad (10.14)$$

If the object is at infinity, the object vergence $L_s = 0$ D. Equation (10.14) then reduces to

$$SM = \frac{1}{1 - dF_{sp}}$$

which is none else but equation (10.8).

## 10.8 Spectacle magnification is magnifying power

Spectacle magnification may be likened to the magnifying power of a telescopic system. When a telescope is adjusted for infinity, the second principal focus of its objective (first lens) and the first principal focus of its eye piece (second lens) must coincide (*Figure 10.10*). This is very

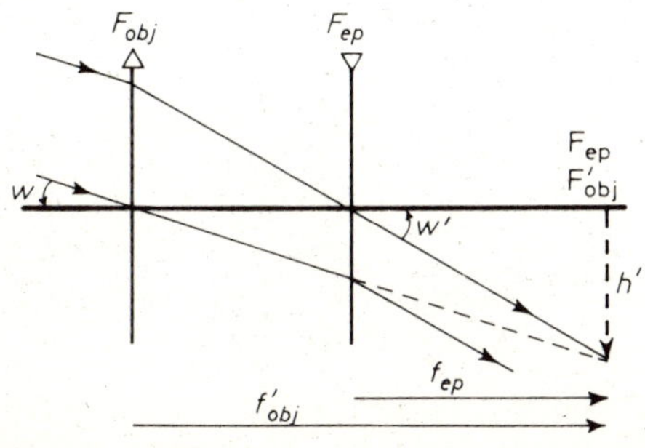


*Figure 10.10 Diagram of a simple telescopic system. ep = eye piece; obj = objective*

similar to the requirements of a spectacle correction for distance vision: here, the second principal focus of the spectacle lens (first lens) must coincide with the far point of the eye. The far point distance may be likened to the focal length of the ocular correction (second lens). The ocular correction could be represented as a contact lens of power $-K$ dioptres placed in touch with an emmetropic eye thus producing an ametropia of $K$ dioptres ocular correction (*Figure 10.11*). The objective of this 'telescopic system' has a power of $F_{sp}$ dioptres, its eye piece

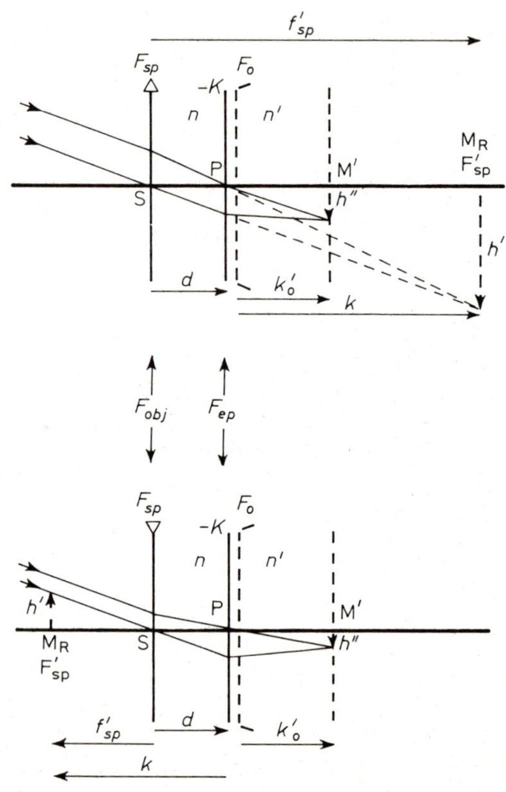


*Figure 10.11 The corrected ametropic eye as a telescopic system*

has a power of $K$ dioptres and the distance between the two components is equal to the vertex distance $d$. The angular magnification or magnifying power of a telescope is given by (*Figure 10.10*)

$$M = \frac{\tan w'}{\tan w}$$

$$= \frac{-h'}{f_{ep}} \Big/ \frac{-h'}{f'_{obj}}$$

$$= \frac{-h'}{f_{ep}} \times \frac{f'_{obj}}{-h'}$$

$$= \frac{f'_{obj}}{f_{ep}}$$

$$= \frac{f'_{obj}}{-f'_{ep}}$$

$$= \frac{-F_{ep}}{F_{obj}}$$

where the subscripts *obj* stand for objective and *ep* for eye piece. In the corrected ametropic eye, $F_{sp} = F_{obj}$ and $-K = F_{ep}$ and so the equation becomes

$$M = \frac{K}{F_{sp}}$$

which, according to equation (10.6), gives spectacle magnification.

This matter has been discussed by Reiner (1965a, 1972) and by Obstfeld (1976b).

# Chapter Eleven
# Relative Spectacle Magnification

## 11.1 Definition

Relative spectacle magnification may be defined as the ratio of the retinal image size of a distant object in the corrected ametropic eye to the retinal image size of that object in the standard reduced emmetropic eye. As a 'formula', this may be written as

$$RSM = \frac{\text{the retinal image size of a distant object in the corrected eye}}{\text{the retinal image size of that object in the standard reduced emmetropic eye}}$$

When formulae have been forgotten, the retinal image sizes in these two cases could be calculated; their ratio provides the relative spectacle magnification.

## 11.2 Introduction

Relative spectacle magnification is mainly associated with binocular vision, in particular with the conditions of anisometropia and aphakia.

Anisometropia (an = not; isos = equal; unequal refractive errors) is the condition of a pair of eyes where the one eye has a markedly different ametropia as compared with the other. For example, the right eye has an ocular correction of +3.00 D and the left eye of +6.00 D. A particular case of anisometropia known as antimetropia, is where the one eye is myopic and its fellow hyperopic.

Aphakia (chapter 14) is the condition where the crystalline lens is absent. This, usually, results in a high degree of hyperopia. In addition, such an eye cannot accommodate. If only one of a pair of eyes is aphakic (unilateral aphakia), particular problems arise with respect to binocular vision.

In these cases it is very well possible that there is a retinal image size difference in a pair of eyes (aniseikonia). When this difference is 5% or more (Baumann, 1966; Trotter, 1967) difficulties may arise with fusion (Le Grand, 1964, says that symptoms may appear when there is a difference of only ½%, while Michaels (1975) quotes a figure as small as ¼%. Besides, Binkhorst and Loones (1975) reported a possible adaptation (reduction) to aniseikonia). Fusion may be defined as the brain faculty whereby one becomes aware of a single impression of external space formed on the basis of the retinal images of each of a pair of eyes.

Only the retinal image size difference in one eye, for different spectacle lens corrections at different vertex distances, can be calculated accurately (spectacle magnification). The retinal image sizes in different eyes cannot usually be calculated accurately because details about the cause of the ametropia are not normally known. If the one eye's ametropia were refractive and that of the other axial, different basic retinal image sizes would be present (chapter 7). The spectacle correction of each eye would increase or decrease any retinal image size difference depending on the type(s) of ametropia and the type of correction.

In order to obtain some idea of the relative differences the retinal image sizes may be compared with a standard, namely the retinal image size of the standard emmetropic reduced eye (Emsley, 1955; Michaels, 1975). Other authors (Bennett, 1956; Siebeck, 1960) used a Gullstrand schematic eye as standard. The ratio of the relative spectacle magnification of the two eyes now indicates the ratio of their retinal image sizes after correction. When an accurate calculation is called for account ought to be taken of the shape factor (section 14.4).

Ametropia has been classified as being either refractive or axial; there is a relative spectacle magnification formula for each case. However, this classification is too stringent. It has been found that anisometropia is due mainly to axial ametropia, but this is usually associated with a smaller difference in the refraction of the two eyes (Sorsby *et al.*, 1962). Refractive ametropia may be caused by drugs and can be present in diabetic myopia and spasm of accommodation. Aphakia is another type. Keratometer measurements may indicate differences in the powers of a pair of eyes. Measurements of the axial length of eyes are difficult to obtain (ultrasonography).

The information obtained from the formulae should, therefore, only be regarded as a pointer and certainly not as the absolute situation.

## 11.3 Derivation of formulae

Several formulae for the relative spectacle magnification will be derived.

Different formulae apply to axial and refractive ametropia while a formula giving relative spectacle magnification in an ametropic eye suffering from both, will be given too.

### 11.3.1 AXIAL AMETROPIA

Equation (10.13) gives the retinal image size $h''_{am}$ for a distant object in a corrected ametropic eye. Thus

$$h''_{am} = \frac{\tan w_o}{F}$$

The retinal image size $h'_{em}$ for a distant object in a standard reduced emmetropic eye is given by (equation (4.3))

$$h'_{em} = -k'_o \frac{3}{4} \tan w_o$$

or

$$\begin{aligned} h'_{em} &= -f'_o \frac{3}{4} \tan w_o \\ &= \frac{-n'}{F_o} \frac{3}{4} \tan w_o \\ &= \frac{-\tan w_o}{F_o} \end{aligned}$$

By definition,

$$RSM_{ax} = \frac{h''_{am}}{h'_{em}} \tag{11.1}$$

Hence,

$$\begin{aligned} RSM_{ax} &= \frac{\tan w_o/F}{\tan w_o/F_o} \\ &= \frac{F_o}{F} \end{aligned} \tag{11.2}$$

For axial ametropia, and using equation (1.8) (*see* section 10.5.4), this may be expanded to read

$$RSM_{ax} = \frac{F_o}{F_{sp} + F_o - dF_{sp}F_o} \tag{11.3}$$

There is another manner in which this concept may be looked at. As shown the ratio of the retinal image sizes in the eyes depends on their respective focal lengths. In eyes where the retinal images are sharp it is true to say that their retinal image sizes depend on their respective axial lengths with the proviso that in the corrected ametropic eye the effect of the spectacle magnification must be taken into account. *See Figure 11.1.*

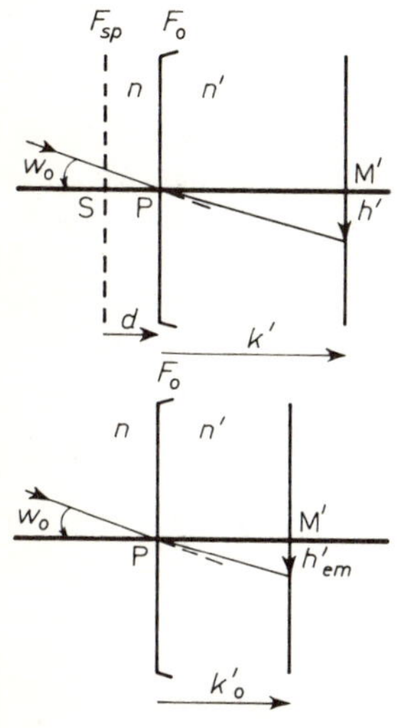


*Figure 11.1* ***Relative spectacle magnification in axially ametropia depends on axial length and spectacle magnification***

The retinal image size $h''_{am}$ of a distant object in a corrected ametropic eye is given by the product of the basic retinal image size $h'$ and the spectacle magnification; thus

$$h''_{am} = h' \times SM$$

In view of equation (4.3),

$$h''_{am} = -k' \frac{3}{4} \tan w_o \times SM$$

Similarly, the retinal image size $h'_{em}$ of the distant object in the standard reduced emmetropic eye is given by (equation (4.3))

$$h'_{am} = -k'_o \frac{3}{4} \tan w_o$$

Substituting into equation (11.1),

$$RSM_{ax} = \frac{k' \frac{3}{4} \tan w_o \times SM}{k'_o \frac{3}{4} \tan w_o}$$

$$= \frac{k'}{k'_o} \times SM \qquad (11.4)$$

This equation may be altered to give the derived equations

(equation (10.5)) $$RSM_{ax} = \frac{k'}{k'_o} \times \frac{f'_{sp}}{k} \qquad (11.5)$$

(equation (10.6)) $$RSM_{ax} = \frac{K'_o}{K'} \times \frac{K}{F_{sp}} \qquad (11.6)$$

### 11.3.2 REFRACTIVE AMETROPIA

In refractive ametropia the ametropic eye is of standard length but the power departs from that of the standard reduced emmetropic eye. Consequently, in equation (11.3) $F_o$ should be replaced by $F_e$ (in the denominator). Thus, the expression will read

$$RSM_r = \frac{F_o}{F_{sp} + F_e - dF_{sp}F_e} \qquad (11.7)$$

Similarly, in equation (11.4) alterations must be made. Since the axial length of both the ametropic and emmetropic eye are the same, the equation simplifies to

$$RSM_r = SM \qquad (11.8)$$

Hence, any one of the equations for spectacle magnification given in the previous chapter, may be employed to calculate the relative spectacle magnification in case of refractive ametropia!

### 11.3.3 MIXED AMETROPIA

If an ametropic eye is subject to both axial and refractive ametropia, the relative spectacle magnification may be calculated too. Equation (11.4) appears to provide the best 'descriptive' solution in such a case.

## 11.4 Calculations

### 11.4.1 AXIAL AMETROPIA

EXAMPLE 17

Calculate the relative spectacle magnification for an axially myopic eye of 10.00 D ocular correction. The vertex distance is 12 mm.

*Solution 17*

In order to be able to calculate the relative spectacle magnification, the spectacle correction must be known. This may be calculated as set out in section 9.2.2. It will show $F_{sp} = -11.36$ D. Substituting these values into equation (11.3), one obtains

$$
\begin{aligned}
RSM_{ax} &= \frac{+60.00}{-11.36 + 60.00 - 0.012(-11.36)(+60.00)} \\
&= \frac{+60.00}{+48.64 + 8.18} \\
&= \frac{+60.00}{+56.82} \\
&= +1.056\,3
\end{aligned}
$$

Alternatively, using equation (11.6):
Calculate the dioptric length of the eye (using the ametropia formula):

$$
\begin{aligned}
K' &= K + F_o \\
&= -10.00 + 60.00 \\
&= +50.00 \text{ D}
\end{aligned}
$$

Now all details required, are known. Hence, by substitution into equation (11.6)

$$
\begin{aligned}
RSM_{ax} &= \frac{+60.00(-10.00)}{+50.00(-11.36)} \\
&= +1.056\,3
\end{aligned}
$$

This means that the retinal image size in this corrected myopic eye is 5.6% larger than that of the same distant object in the standard reduced emmetropic eye.

EXAMPLE 18

Calculate the relative spectacle magnification for an axially hyperopic eye of ocular correction +10.00 D. The vertex distance is 12 mm.

*Solution 18*

This problem may be solved in the same ways as example 17. The spectacle correction will be found to be +8.93 D. The relative spectacle magnification is then found to be +0.960 0.

This means that the retinal image size in the corrected hyperopic eye is 4% smaller than that in the standard reduced emmetropic eye.

When considering the value found for the axially myopic eye it has to be remembered that such an eye is axially longer than the standard reduced emmetropic eye (*Figure 7.2*), and that the retinal image will, therefore, be larger notwithstanding the reduction ('minification') caused by the spectacle lens. The opposite is true for the hyperopic eye.

### 11.4.2 REFRACTIVE AMETROPIA

EXAMPLE 19

Calculate the relative spectacle magnification for a case of refractive ametropia of ocular correction −10.00 D. Vertex distance 12 mm.

*Solution 19(a)*

The power of the eye may be calculated from the ametropia formula:

$$
\begin{aligned}
F_e &= K'_o - K \\
&= +60.00 - (-10.00) \\
&= +70.00 \text{ D}
\end{aligned}
$$

The spectacle correction is −11.36 D (*see* example 17). By substitution into equation (11.7),

$$
\begin{aligned}
RSM_r &= \frac{+60.00}{-11.36 + 70.00 - 0.012(-11.36)(+70.00)} \\
&= \frac{+60.00}{+58.64 + 9.54} \\
&= \frac{+60.00}{+68.18} \\
&= +0.880\ 0
\end{aligned}
$$

This means that the retinal image size of a distant object in the corrected myopic eye is 12% smaller than that in the standard reduced emmetropic eye.

EXAMPLE 20

Calculate the relative spectacle magnification for an eye having refractive ametropia, of ocular correction +10.00 D. Vertex distance 12 mm.

*Solution 20(a)*

The spectacle correction is +8.93 D (see also example 18), and the power of the eye will be found to be +50.00 D. The calculation will show that (equation (11.7))

$$
\begin{aligned}
RSM_r &= \frac{+60.00}{+8.93 + 50.00 - 0.012(+8.93)(+50.00)} \\
&= \frac{+60.00}{+58.93 - 5.36} \\
&= \frac{+60.00}{+53.57} \\
&= +1.120\,0
\end{aligned}
$$

which indicates that the retinal image of the corrected hyperopic eye is 12% larger than that of the standard reduced emmetropic eye, for the same distant object.

With reference to equation (11.8), the following calculations will provide proof that the relative spectacle magnification and spectacle magnification for cases of refractive ametropia give the same value.

*Solution 19(b)*

By substitution into equation (10.6)

$$
\begin{aligned}
RSM_r &= \frac{-10.00}{-11.36} \\
&= +0.880\,3
\end{aligned}
$$

*Solution 20(b)*

By substitution into equation (10.6)

$$
\begin{aligned}
RSM_r &= \frac{+10.00}{+8.93} \\
&= +1.119\,8
\end{aligned}
$$

## 11.5 Evaluation

On comparing the values calculated in the previous section, for the relative spectacle magnification of axial and refractive ametropia, the reader will notice that there is a large difference:

1. Since relative spectacle magnification for high degrees of axial ametropia does not depart a great deal from unity, medium and low degrees of axial ametropia will show relative spectacle magnification values very close to unity. This is due to the fact that the basic retinal image size difference between ametropic and emmetropic eyes is counteracted by the spectacle magnification produced by the correction of the ametropic eye.
2. Since relative spectacle magnification for high degrees of refractive ametropia departs a great deal from unity, medium and low degrees will show a proportionally large departure from unity. Since the basic retinal image sizes in refractively ametropic and emmetropic eyes are the same, it is caused purely by the spectacle magnification of the correction of the ametropic eye.

It is, therefore, of interest to consider briefly the correction of cases of refractive ametropia with contact lenses. Since the contact lens is fitted on the cornea of a (real) eye, it will have a very short 'vertex distance' (to the entrance pupil of the eye; *see* section 20.3). This will be of the order of 3 mm. Hence, the difference between ocular correction and 'spectacle correction' (that is, contact lens correction, in this context) will be almost negligible. Relative spectacle magnification and, incidentally, spectacle magnification, will depart only very slightly from unity.

These observations are summarized in *Table 11.1*, which shows also the effect of spectacle magnification.

Under certain conditions relative spectacle magnification assumes unit value for other than emmetropic eyes. These conditions were given by Knapp (1870) and are known as Knapp's Rule or Knapp's Law. They are the following:

1. The ametropia must be axial.
2. The power of the eye must be equal to that of the standard emmetropic eye.
3. The correcting lens must be placed at the anterior focus of the eye.
4. The shape factor is of unit value.

TABLE 11.1

| | *Hyp* | *Emm* | *Myo* |
|---|---|---|---|
| *SM* | | | |
| value | more than +1 | +1 | less than +1 |
| when *d* larger, value | increased | +1 | decreased |
| $RSM_{ax}$ | | | |
| basic retinal image size | smaller than Emm | | bigger than Emm |
| sharp retinal image size | nearly Emm size | | slightly bigger than Emm size |
| value | slightly less than +1 | +1 | slightly more than +1 |
| $RSM_r$ | | | |
| basic retinal image size | Emm size | | Emm size |
| sharp retinal image size | larger than Emm size | | smaller than Emm size |
| value | more than +1 | +1 | less than +1 |

A graphical representation of these observations was given by Bennett (1956)

NOTE
Knapp's Rule does not apply to spectacle magnification.

Under these conditions the denominator in equation (11.2) assumes the same value as the numerator. Hence, the relative spectacle magnification becomes of unit value.

## 11.6 Binocular vision

The ratio of the relative spectacle magnification of each of a pair of eyes indicates the ratio of their retinal image sizes (after correction). For the purpose of an accurate calculation, account should be taken of the shape factor of each correcting lens though this is usually omitted.

The retinal image size ratio (*RISR*) of a pair of eyes is, therefore, given by

$$RISR = \frac{RSM_R}{RSM_L} \tag{11.9}$$

where the suffixes indicate the value for the right and left eye respectively. Any one of the formulae for relative spectacle magnification may be used, as appropriate to the circumstances.

Consider the following case of anisometropia:

EXAMPLE 21
The spectacle corrections of a pair of axially ametropic eyes are R + 1.00 D and L + 5.00 D. Vertex distance 12 mm. Calculate the retinal image size ratio.

*Solution 21*
For the right eye:

$$RSM_R = \frac{+60.00}{+1.00 + 60.00 - 0.012(+1.00)(+60.00)}$$

$$= \frac{+60.00}{+61.00 - 0.72}$$

$$= \frac{+60.00}{+60.28}$$

$$= +0.995\,4$$

For the left eye:

$$RSM_L = \frac{+60.00}{+5.00 + 60.00 - 0.012(+5.00)(+60.00)}$$

$$= \frac{+60.00}{+65.00 - 3.60}$$

$$= \frac{+60.00}{+61.40}$$

$$= +0.977\,2$$

Hence (equation (11.9)),

$$RISR = \frac{+0.995\,4}{+0.977\,2}$$

$$= +1.018\,6$$

which is indicative of a retinal image size difference between the two eyes of almost 2%. This is not likely to cause difficulties with binocular vision, since the difference is less than 5%.

EXAMPLE 22
The spectacle correction of a pair of eyes known to have refractive ametropia, is R + 1.00 and L + 5.00 D. Vertex distance 12 mm. Comment on the likely state of binocular vision.

*Solution 22*
For the right eye:
Since any spectacle magnification formula can be used to determine the

relative spectacle magnification of this eye, equation (10.8) would be convenient in view of the data provided. Hence

$$
\begin{aligned}
RSM_R &= \frac{1}{1-0.012(+1.00)} \\
&= \frac{1}{0.988} \\
&= +1.012\,1
\end{aligned}
$$

Similarly, for the left eye:

$$
\begin{aligned}
RSM_L &= \frac{1}{1-0.012(+5.00)} \\
&= \frac{1}{1-0.06} \\
&= \frac{1}{0.94} \\
&= +1.063\,8
\end{aligned}
$$

The retinal image size ratio is

$$
\begin{aligned}
RISR &= \frac{+1.012\,1}{+1.063\,8} \\
&= +0.951\,4
\end{aligned}
$$

which indicates that there would be an almost 5% difference in the retinal image sizes of these eyes.

COMMENT

Since this value has been generally accepted to represent the border between binocular vision with and without difficulties, the wearer may or may not be free from symptoms. It would, therefore, be advisable to try to reduce the vertex distance. This will reduce the difference between ocular and spectacle correction, and hence the disparity between the retinal image sizes will be reduced. This may result in symptom-free binocular vision.

The reader will appreciate that, instead of carrying out two separate calculations to obtain the relative spectacle magnification for each eye, formulae can be compiled which will incorporate these relative spectacle

magnifications to give the retinal image size ratio directly. The following formulae have regard to equation (11.9), and the suffixes $R$ and $L$ indicate a value for the right or left eye respectively.

From (11.2) $$RISR_{ax} = \frac{F_L}{F_R} \tag{11.10}$$

From (11.3) $$RISR_{ax} = \frac{F_{spL} + F_o - d_L F_{spL} F_o}{F_{spR} + F_o - d_R F_{spR} F_o} \tag{11.11}$$

From (11.4) $$RISR_{ax} = \frac{k'_L SM_L}{k'_R SM_R} \tag{11.12}$$

From (11.5) $$RISR_{ax} = \frac{k_L k'_R f'_{spR}}{k_R k'_L f'_{spL}} \tag{11.13}$$

From (11.6) $$RISR_{ax} = \frac{K_R K'_L F_{spL}}{K_L K'_R F_{spR}} \tag{11.14}$$

From (11.7) $$RISR_r = \frac{F_{spL} + F_{eL} - d_L F_{spL} F_{eL}}{F_{spR} + F_{eR} - d_R F_{spR} F_{eR}} \tag{11.15}$$

From (10.5) $$RISR_r = \frac{k_L f'_{spR}}{k_R f'_{spL}} \tag{11.16}$$

From (10.6) $$RISR_r = \frac{K_R F_{spL}}{K_L F_{spR}} \tag{11.17}$$

From (10.8) $$RISR_r = \frac{1 - d_L F_{spL}}{1 - d_R F_{spR}} \tag{11.18}$$

From (10.11) $$RISR_r = \frac{1 + d_R K_R}{1 + d_L K_L} \tag{11.19}$$

## 11.7 Use in calculations

Since relative spectacle magnification provides the ratio of the corrected ametropic eye's retinal image size and that of the standard reduced emmetropic eye (for distant objects), it may be employed to calculate the retinal image size in the corrected ametropic eye on the basis of the retinal image size of the emmetropic eye. For the purpose of an example, reconsider example 16 (section 10.5). The problem reads as follows: A distant object subtends an angle $w_o$ at an axially ametropic reduced

eye of ocular correction +5.00 D. This eye is corrected by a spectacle lens at 15 mm. Calculate the retinal image size.

*Solution 16*
*Step 1.* The retinal image size in the standard reduced emmetropic eye is given by equation (4.3):

$$h' = -k'_o \frac{3}{4} \tan w_o$$

or

$$h' = -22.22 \times \frac{3}{4} \tan w_o$$

$$= -16.67 \tan w_o \text{ mm}$$

*Step 2.* From earlier calculations, the spectacle correction was found to be +4.65 D.
*Step 3.* The relative spectacle magnification may be calculated from equation (11.3):

$$RSM_{ax} = \frac{+60.00}{+4.65 + 60.00 - 0.015(+4.65)(+60.00)}$$

$$= \frac{+60.00}{+64.65 - 4.19}$$

$$= \frac{+60.00}{+60.46}$$

$$= +0.992\,4$$

*Step 4.* Hence the retinal image size in the corrected ametropic eye is given by

$$h'' = RSM_{ax} \times h'$$

$$= +0.992\,4(-16.67 \tan w_o) \text{ mm}$$

$$= -16.54 \tan w_o \text{ mm}$$

This compares well with the values calculated earlier.

## 11.8 Universal formula for relative spectacle magnification

Relative spectacle magnification has been defined with respect to a distant object. However, it is of theoretical interest to know that a Universal Relative Spectacle Magnification formula may be developed by

reasoning along lines similar to those discussed in section 10.7 for spectacle magnification.

Hence, for axial ametropia, equation (10.14) may be substituted into equation (11.4) to read

$$RSM_{ax} = \frac{k'}{k'_o} \times \frac{1 - dL_s}{1 - dL'_s} \tag{11.20}$$

Relative spectacle magnification in cases of refractive ametropia has been shown to be equal in value to spectacle magnification. Hence, the Universal Relative Spectacle Magnification for refractive ametropia is given by equation (10.14).

# Chapter Twelve
# Accommodation

## 12.1 Definition

Accommodation may be defined as the faculty of the eye to increase its dioptric power so that an object situated nearer by than the far point may be imaged on the retina.

## 12.2 Introduction

It is outside the scope of this text to deal with this physiological process in any detail. Any textbook of ocular physiology may be consulted to obtain information about accommodation. Let it, therefore, suffice to say that when the eye accommodates the radii of curvature of the crystalline lens are shortened. This results in an increase in the power of the lens. In turn, the total power of the eye is increased.

## 12.3 Ocular accommodation

It should be pointed out that in the following sections reference will be made only to 'ocular accommodation'. This may be defined as the accommodation exerted by the (reduced) eye as measured at the (reduced) surface. 'Spectacle accommodation' will be discussed in section 12.7.

## 12.4 Visual optics

In the reduced eye the amount of accommodation exerted by the eye, indicated by the symbol $A$ and measured in dioptres, is added to the power of the eye in its unaccommodated (or relaxed) state. Thus, if

an eye of power $F_e$ = +62.00 D accommodates 5.00 D, the power of the accommodated eye is given by

$$F_a = F_e + A \tag{12.1}$$

$$= +62.00 + 5.00$$

$$= +67.00 \text{ D}$$

If this eye had been of standard power, the equation would read

$$F_a = F_o + A \tag{12.2}$$

What are the consequences of accommodation? Assuming that the +62.00 D eye is of standard axial length, the ametropia formula shows an ocular correction of

$$K = K_o' - F_e$$

$$= +60.00 - (+62.00)$$

$$= -2.00 \text{ D}$$

The far point distance is thus (equation (5.7))

$$k = \frac{100}{-2.00}$$

$$= -50.00 \text{ cm}$$

The eye now accommodates 5.00 D. Its power increases to $F_a$ = +67.00 D. It will be clear that the far point $M_R$ is no longer conjugate with the retina at M′. But which point is?

Before introducing further symbols associated with accommodation, the conjugate foci formula (equation (1.12)) may be used. By substituting the power of the accommodated eye $F_a$ for the power of the optical system $F$ and the (standard) dioptric length of the eye $K_o'$, for the image vergence $L'$ we get

$$K_o' = L + F_a$$

Hence, by substituting the appropriate values, the object vergence is

$$L = +60.00 - (+67.00)$$

$$= -7.00 \text{ D}$$

The object distance is then given by equation (1.10)

$$l = \frac{100}{-7.00}$$

$$= -14.29 \text{ cm}$$

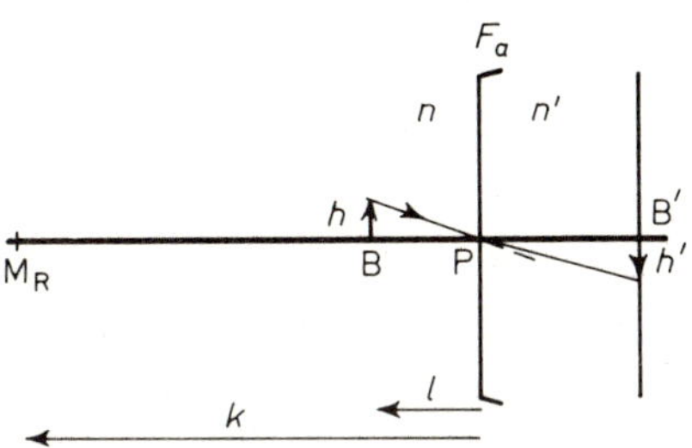


*Figure 12.1 As a result of accommodation, the object plane (at* B*) is conjugate with the retina (at* B'*)*

This means that a point 14.29 cm in front of the eye is now conjugate with the retina of that eye (*Figure 12.1*).

## 12.5 Accommodation formulae

### 12.5.1 GENERAL FORMULA

It is not normally necessary to carry out the calculation as set out in the previous section, that is, with respect to the retina or the dioptric length of the eye. It is simpler to calculate the data required with respect to the reduced surface of the eye, as follows.

The most general of the ametropia formulae, reads (equation (5.6))

$$K' = K + F_e$$

In case of accommodation, the object under observation will be placed at a point nearer than the far point, say, at a distance of $l$ metres ($l < k$). The object vergence with respect to the reduced surface is $L$ dioptres. This object is imaged on the retina and so the formula reads

$$K' = L + F_a$$

In view of equation (5.6),

$$K + F_e = L + F_a$$

This, in turn, may be modified in view of equation (12.1), thus

$$K + F_e = L + F_e + A$$

which may be simplified and written as

$$A = K - L \qquad (12.3)$$

NOTE

No details about the cause of the eye's ametropia are required.

EXAMPLE 23

A 5.00 D myope sees clearly an object situated 10 cm in front of his eye. Calculate the accommodation required.

*Solution 23*

*Step 1.* Draw a diagram (*Figure 12.1*).

*Step 2.* The object vergence is $L = -10.00$ D.

*Step 3.* Substitution into equation (12.3) provides the answer:

$$A = -5.00 - (-10.00)$$
$$= +5.00 \text{ D}$$

EXAMPLE 24

A hyperope of 5.00 D sees clearly an object at infinity. Calculate the accommodation required.

*Solution 24*

*Step 1.* The object vergence is $L = 0.00$ D.

*Step 2.* Substitution into equation (12.3) shows

$$A = +5.00 - 0.00$$
$$= +5.00 \text{ D}$$

NOTE

This calculation shows that an uncorrected hyperope has to accommodate an amount equal to his ametropia, in order to see a distant object clearly. This was to be expected since ametropia is defined with respect to infinity.

EXAMPLE 25

A 5.00 D hyperope can see clearly an object 50 cm in front of his eye. Calculate the accommodation required.

*Solution 25*

*Step 1.* Draw a diagram (*Figure 12.2*).

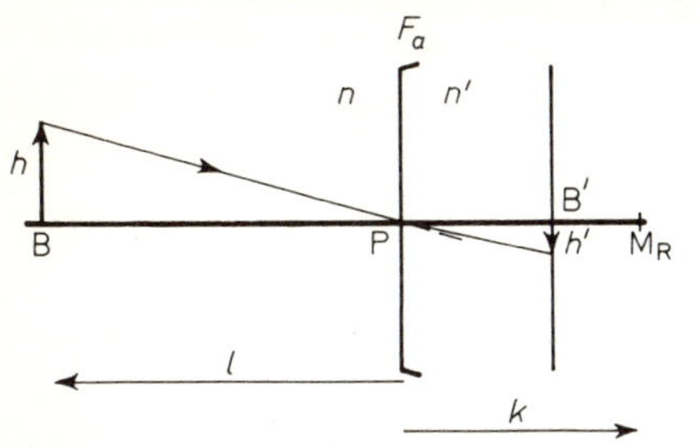


*Figure 12.2 Example 25*

*Step 2.* The object vergence is $L = -2.00$ D.

*Step 3.* Substitution into equation (12.3) gives

$$A = +5.00 - (-2.00)$$
$$= +7.00 \text{ D}$$

### 12.5.2 DONDERS' FORMULA

The nearest point for which an eye can accommodate is known as the near point (of accommodation). It is indicated by $M_P$ where the suffix P stands for 'proximum'; hence, it is also known as the punctum proximum. The near point distance is indicated by $b$ (= $PM_P$) and the near point vergence $B$ is given by

$$B = \frac{n}{b} \tag{12.4}$$

The near point is associated with the maximum amount of accommodation that an eye can exert: the absolute amplitude of accommodation. (The maximum amount of accommodation exerted by the pair of eyes is known as the binocular amplitude of accommodation.) The amplitude of accommodation is indicated by *Amp* and is measured in dioptres. Donders' formula (1864), which deals with the amplitude of accommodation, may be derived by substituting the near point vergence $B$ for the object vergence $L$ of equation (12.3), thus

$$Amp = K - B \tag{12.5}$$

EXAMPLE 26

An eye's far point is at −40 cm. It has am amplitude of accommodation of 4.50 D. Calculate the near point distance.

*Solution 26*
*Step 1.* Draw a diagram (*Figure 12.3*).

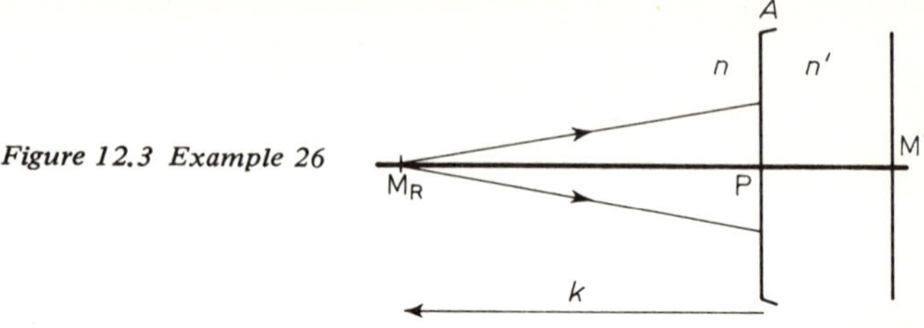

*Figure 12.3 Example 26*

*Step 2.* The ametropia is $K = -2.50$ D.
*Step 3.* Substituting the data into equation (12.5), the near point vergence is

$$B = -2.50-(+4.50)$$
$$= -7.00 \text{ D}$$

*Step 4.* The near point distance is (equation (12.4))

$$b = \frac{100}{-7.00}$$
$$= -14.29 \text{ cm}$$

### 12.5.3 CONJUGATE FOCI

If, for the sake of argument, accommodation were provided in the form of an imaginary contact lens of power $A$, fitted in the plane of the reduced surface, it is true to say that the near object, when imaged on the retina, is also conjugate with the far point through this imaginary contact lens. Thus equation (12.3) may be rearranged for the purpose of resembling the conjugate foci formula (1.12) by substituting $A$ for the focal power and $K$ for the image vergence. It then follows that

$$K = L+A$$

Having regard to the principle of the reversibility of the ray path, $L$ may also be taken as the image vergence and $K$ as the object vergence. Hence,

$$L = K-A$$

where $A$ must be subtracted since the direction of travel of the rays has been reversed.

The same reasoning may be applied to Donders' formula (equation (12.
Hence

$$K = B + Amp$$

and

$$B = K - Amp$$

## 12.6 Associated changes

During accommodation the power of the eye increases by the number of dioptres accommodated. As a result of the change in power, the focal lengths and the position of the nodal point change too. Hence, the following relationships hold good:

From equation (1.8) $$f_a = \frac{-n}{F_a} \qquad (12.6)$$

From equation (1.9) $$f_a' = \frac{n'}{F_a} \qquad (12.7)$$

From equation (1.4) $$r_a = \frac{n' - n}{F_a} \qquad (12.8)$$

## 12.7 Retinal image size – uncorrected eye

When an object is placed at a distance $l$ ($l < k$) and the eye can accommodate for that distance, the magnification of the optical system is given by

$$m = \frac{L}{K'} \qquad (12.9)$$

which is only a modification of equation (6.3), where $K$ has been substituted by $L$.

In the case of a reduced eye of standard length, equation (12.9) is altered to read

$$m = \frac{L}{K_o'} \qquad (12.10)$$

NOTE

The amount of accommodation exerted need not be known in these cases.

The actual retinal image size may be calculated from equation (1.13):

$$h' = mh$$

Alternatively, the geometrical method, explained in section 6.2, may be employed.

EXAMPLE 27

An accommodated reduced eye of standard length can see clearly a 2 mm high object situated 20 cm in front of it. Calculate the retinal image size.

*Solution 27*

According to equation (12.9):

*Step 1.* Draw a diagram (*Figure 12.4*).

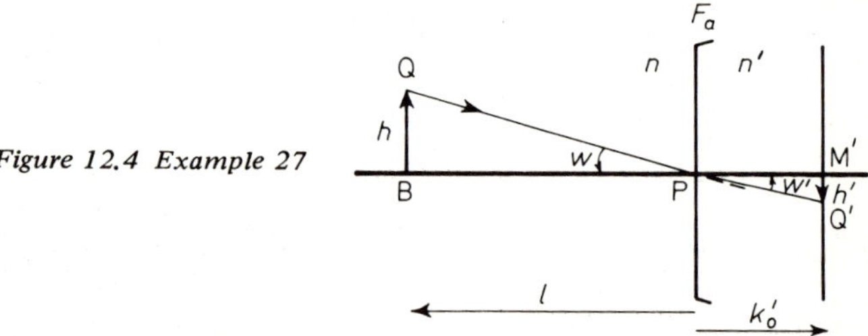


*Figure 12.4 Example 27*

*Step 2.* The object vergence is $L = -5.00$ D.

*Step 3.* By substitution into equation (12.10), the magnification is found to be

$$m = \frac{-5.00}{+60.00}$$
$$= -0.083\ 3$$

*Step 4.* The retinal image size may now be calculated from equation (1.13):

$$h' = -0.083\ 3 \times 2$$
$$= -0.17 \text{ mm}$$

According to the geometrical method:

*Step 1.* Draw a diagram (*Figure 12.4*).

*Step 2.* Consider ΔΔBQP and M′Q′P:

$$\tan w' = \frac{3}{4}\tan w$$

or

$$\frac{-h'}{k_o'} = \frac{3}{4} \times \frac{h}{-l}$$

Thus

$$\begin{aligned} h' &= -\frac{3}{4} \times \frac{hk_o'}{-l} \\ &= \frac{3 \times 2 \times 22.22}{4(-200)} \\ &= -0.17 \text{ mm} \end{aligned}$$

The reader is advised to consider problems encountered when the object distance has a positive value.

## 12.8 Spectacle accommodation

This may be defined as the amount of acommodation required (or exerted) as measured at the spectacle point, to see clearly an object placed in front of the spectacle lens. This implies that the ametropia has been corrected, that is, a distance correction is being worn. Consequently, the spectacle accommodation $A_s$ needs to 'neutralize' only the object vergence $L_s$ with respect to the spectacle point.
Hence,

$$A_s = -L_s \tag{12.11}$$

EXAMPLE 28
Calculate the spectacle accommodation required to see clearly an object placed 25 cm in front of a corrected eye. The vertex distance is 12 mm.

*Solution 28*
*Step 1.* Draw a diagram (*Figure 12.5*).
*Step 2.* Calculate the object distance $l_s$ with respect to the spectacle point:

$$\begin{aligned} l_s &= l + d \\ &= -25.00 + (+1.2) \\ &= -23.80 \text{ cm} \end{aligned}$$

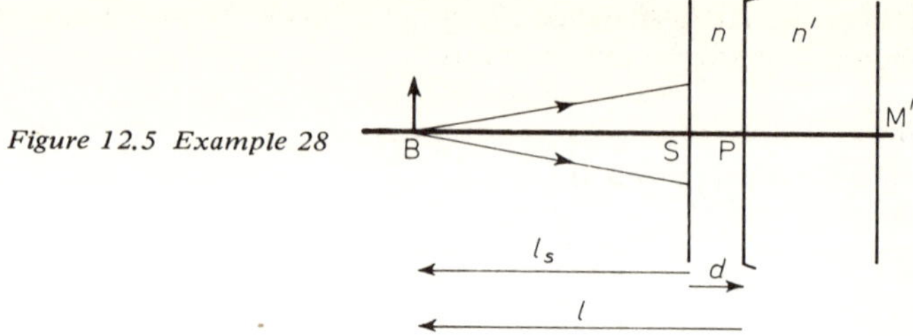


*Figure 12.5 Example 28*

*Step 3.* Calculate the object vergence with respect to the spectacle point:

$$L_s = \frac{100}{-23.80}$$

$$= -4.20 \text{ D}$$

*Step 4.* Calculate the spectacle accommodation from equation (12.11):

$$A_s = -(-4.20)$$

$$= +4.20 \text{ D}$$

## 12.9 Accommodation in the corrected eye

The corrected ametropic eye is normally corrected for distance vision. The object vergence is then assumed to be zero dioptre. For any near object the object vergence $L_s$ will be a negative quantity. Consequently, the image vergence $L_s'$ emerging from the spectacle lens will not be equal to the power of that lens. In turn, the vergence $L$ incident on the eye will not be equal to the ocular correction as is required for the object to be imaged on the retina of the eye. Since the object vergence $L_s$ is negative, the vergence incident on the eye will be less positive or more negative than the ocular correction would demand. The effect of the vertex distance must be taken into account.

Thus, for the object to be imaged on the retina the amount of excess negative power must be 'neutralized'. This is achieved when the eye accommodates.

To calculate how much ocular accommodation is required when an ametropic eye, which is corrected for distance vision, looks at a near object, it is necessary to apply either the 'step-along' method of calculation or a formula derived on the same basis. Both are shown below.

EXAMPLE 29

An object is situated at a distance of 10 cm from the spectacle lens which corrects an eye for distance vision. The power of the lens is

−8.00 D, and the vertex distance 12 mm. Calculate the ocular accommodation required to see the object clearly.

*Solution 29*
*Step 1.* Draw a diagram (*Figure 12.6*).

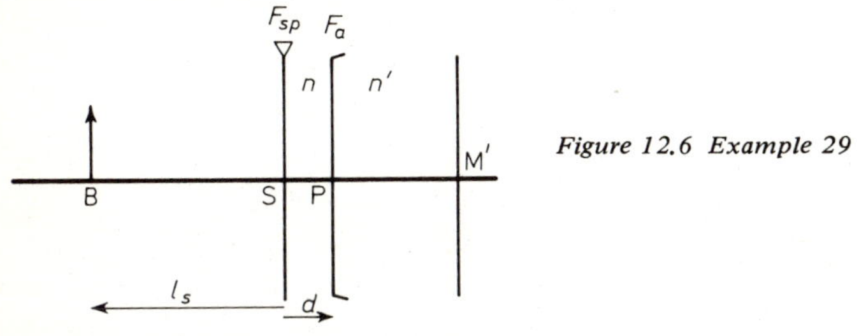


*Figure 12.6 Example 29*

*Step 2.* To calculate the vergence incident on the eye:

$$
\begin{array}{lllcll}
L_s & = & -10.00\text{ D} & \longleftarrow & l_s & = -10.00\text{ cm} \\
F_{sp} & = & -8.00 & + & & \\
\hline
L'_s & = & -18.00 & \longrightarrow & l'_s & = -5.56 \\
 & & & & d & = +1.2 \quad - \\
\hline
L & = & -14.79 & \longleftarrow & l & = -6.76
\end{array}
$$

(For the optical construction, *see Figure 12.7.*)

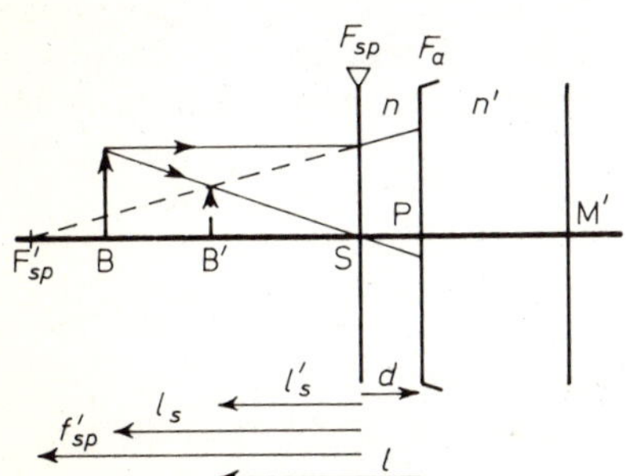


*Figure 12.7 Example 29; the optical construction of the image formed by the spectacle lens*

*Step 3.* Calculate the object vergence that will be focussed on the retina by the unaccommodated eye which is corrected for distance, that is, the ocular correction:

$$
\begin{array}{lllcll}
F_{sp} & = & -8.00\text{ D} & \longrightarrow & f'_{sp} & = -12.50\text{ cm} \\
 & & & & d & = +1.2 \quad - \\
\hline
K & = & -7.30 & \longleftarrow & k & = -13.70
\end{array}
$$

(Alternatively, the effectivity formula (equation (9.4)) may be used.)
*Step 4.* Substitute these data into equation (12.3) to calculate the amount of accommodation required, thus:

$$A = -7.30-(-14.79)$$
$$= +7.49 \text{ D}$$

NOTE

The reader might find it easier to remember Steps 2 and 3 by asking himself:

'What have we got?' (Step 2); and
'What do we want?' (Step 3),

while the difference gives the accommodation required.

The calculation of Step 2 above may be transformed into a formula in the following way:
Assuming that the refractive index of the medium in which the components are placed is 1, the derivation may proceed thus:

$$L_s = \frac{1}{l_s}$$
$$L_s' = L_s + F_{sp}$$

or

$$L_s' = \frac{1}{l_s} + \frac{1}{f_{sp}'}$$
$$= \frac{f_{sp}' + l_s}{l_s f_{sp}'}$$

Hence,

$$l_s' = \frac{l_s f_{sp}'}{f_{sp}' + l_s}$$

Since

$$l = l_s' - d$$
$$= \frac{l_s f_{sp}' - d(f_{sp}' + l_s)}{f_{sp}' + l_s}$$

the vergence incident on the eye is given by

$$\begin{aligned} L &= \frac{f'_{sp} + l_s}{l_s f'_{sp} - d(f'_{sp} + l_s)} \\ &= \left(\frac{1}{F_{sp}} + \frac{1}{L_s}\right) \Big/ \left(\frac{1}{L_s F_{sp}} - \frac{d}{F_{sp}} - \frac{d}{L_s}\right) \\ &= \left(\frac{L_s + F_{sp}}{F_{sp} L_s}\right) \Big/ \left(\frac{1 - d(L_s + F_{sp})}{L_s F_{sp}}\right) \\ &= \left(\frac{L_s + F_{sp}}{F_{sp} L_s}\right) \left(\frac{L_s F_{sp}}{1 - d(L_s + F_{sp})}\right) \\ &= \frac{L_s + F_{sp}}{1 - d(L_s + F_{sp})} \end{aligned} \qquad (12.12)$$

It will be seen by inspection that this formula is only a modification of equation (9.4), the effectivity formula.

When the details are substituted into equation (12.12), the vergence incident on the eye is found to be

$$\begin{aligned} L &= \frac{-10.00 - 8.00}{1 - 0.012(-10.00 - 8.00)} \\ &= \frac{-18.00}{1 + 0.216} \\ &= -14.80 \text{ D} \end{aligned}$$

This calculation is then followed by Steps 3 and 4, as shown above. Next a case of hyperopia will be considered.

EXAMPLE 30

An object is situated at a distance of −10 cm from the spectacle point. The spectacle correction is + 8.00 D. This distance correction is worn at 12 mm from the reduced surface. Calculate the ocular accommodation.

*Solution 30*

*Step 1.* Draw a diagram (*Figure 12.8*).

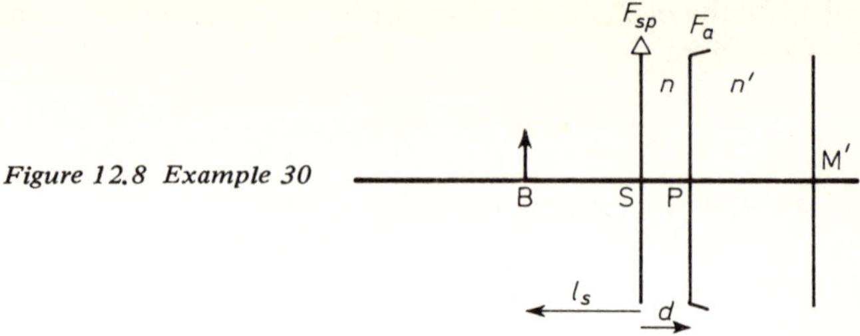


*Figure 12.8 Example 30*

*Step 2.* Calculate the vergence incident on the eye.
Employing the 'step-along' method:

$$
\begin{array}{lll}
L_s = -10.00\text{ D} & \longleftarrow & l_s = -10.00\text{ cm} \\
F_{sp} = +8.00 & + & \\
\hline
L_s' = -2.00 & \longrightarrow & l_s' = -50.00 \\
 & & d = +1.2 \quad - \\
\hline
L = -1.95 & \longleftarrow & l = -51.2
\end{array}
$$

Employing equation (12.12)

$$
\begin{aligned}
L &= \frac{-10.00 + 8.00}{1 - 0.012(-10.00 + 8.00)} \\
&= \frac{-2.00}{1 + 0.024} \\
&= -1.95\text{ D}
\end{aligned}
$$

(*See Figure 12.9* for the optical construction.)

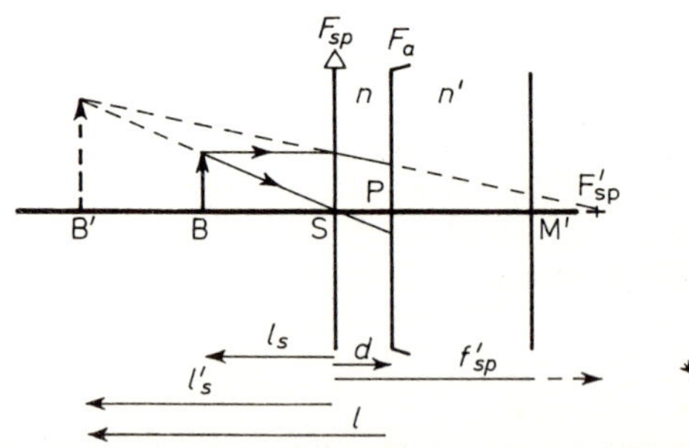


*Figure 12.9 Example 30; the optical construction of the image formed by the spectacle lens*

*Step 3.* Calculate the ocular correction. Either the 'step-along' method or equation (9.4) gives

$$K = +8.85 \text{ D}$$

*Step 4.* Substitution into equation (12.3) produces

$$\begin{aligned} A &= +8.85 - (-1.95) \\ &= +10.80 \text{ D} \end{aligned}$$

Compare this value with that of the previous example (29).

NOTE
A corrected hyperope of given spectacle correction will have to accommodate more to see an object at a certain near vision distance, than a corrected myope having the same spectacle correction (of opposite sign).
This is due to the effectivity at the eye of the vergence emerging from the spectacle lens as well as the difference in vergence of these pencils.

NOTE
When the vertex distance is reduced to (virtually) zero, that is, in the case of a contact lens correction, the hyperope will have to accommodate less than he used to do, but the myope will have to accommodate more. This has implications for the near-presbyopic spectacle wearer who changes from spectacles to a contact lens correction (*see* section 20.9).

## 12.10 Retinal image size – corrected eye

The calculation of the retinal image size in the accommodated corrected eye is similar to that of the corrected eye. The difference is that the object is not situated at infinity.

Consider the following case.

EXAMPLE 31
An axially ametropic eye is corrected by a −6.00 D lens at 10 mm. It sees clearly a 5 mm high object situated 20 cm in front of the spectacle plane. Calculate the retinal image size.

*Solution 31*

*Step 1.* Draw a diagram (*Figure 12.10*).

*Step 2.* Calculate the dioptric length of the eye.

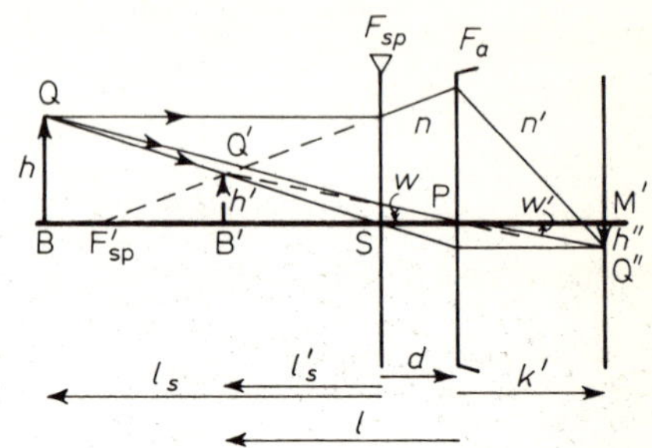


*Figure 12.10 Example 31*

Since this is a case of axial ametropia, the unaccommodated eye must be of standard power. The dioptric length may thus be calculated from the ametropia formula, after calculation of the ocular correction. The latter may be calculated from equation (9.4) or by means of the 'step-along' method. Either will show $K = -5.66$ D. By substitution, the dioptric length is found to be

$$K' = -5.66 + 60.00$$
$$= +54.34 \text{ D}$$

*Step 3.*

NOTE

Since the problem states that the object is seen clearly, it must mean that the 'dioptric length'-vergence is achieved. Since there is no need to calculate the actual ocular accommodation, only the vergence incident on the eye needs to be found. Using the 'step-along' method, this is found to be

| Vergence | | Distance |
|---|---|---|
| $L_s = -5.00$ D | ← | $l_s = -20.00$ cm |
| $F_{sp} = -6.00$ (+) | | |
| $L'_s = -11.00$ | → | $l'_s = -9.09$ |
| | | $d = +1.0$ (−) |
| $L = -9.91$ | ← | $l = -10.09$ |

*Step 4.*
The retinal image size may now be calculated by applying equation (1.15) to each of the two parts of the optical system:

$$m_s = \frac{L_s}{L_s'}$$

$$= \frac{-5.00}{-11.00}$$

$$= +0.454\ 5$$

and

$$m = \frac{L}{K'}$$

$$= \frac{-9.91}{+54.34}$$

$$= -0.182\ 4$$

Hence, the retinal image size is given by

$$h'' = hm_s m$$

which is a double application of equation (1.13).
Thus

$$h'' = 5(+0.454\ 5)(-0.182\ 4)$$

$$= -0.41 \text{ mm}$$

Alternatively, the geometrical method may be used. In that case, Steps 1 and 2 remain the same after which one proceeds as follows.
*Step 3.* Calculate the axial length of the eye (equation (5.1)):

$$k' = \frac{1.333 \times 1\ 000}{+54.34}$$

$$= +24.53 \text{ mm}$$

*Step 4.* Calculate the position of the image $h'$ formed by the spectacle lens of the object $h$. Step 3 of the first solution showed this to be $l_s' = -9.09$ cm.

*Step 5.* Consider ΔΔBQS and B′Q′S:

$$\frac{h'}{h} = \frac{l'_s}{l_s}$$

or

$$h' = \frac{hl'_s}{l_s}$$

By substitution,

$$h' = \frac{5(-9.09)}{-20.00}$$
$$= +2.27 \text{ mm}$$

*Step 6.* Now this image acts as object for the optical system of the eye. Since the image is focussed on the retina, ΔΔB′Q′P and Q″M′P may be considered. Here

$$\tan w' = \frac{3}{4} \tan w$$

which may be written as

$$\frac{-h''}{k'} = \frac{3}{4} \frac{h'}{(-l'_s + d)}$$

Hence,

$$h'' = -\frac{3}{4} \frac{h'k'}{(-l'_s + d)}$$

and by substitution

$$h'' = -\frac{3 \times 2.27 \times 24.53}{4(+90.9 + 10)}$$
$$= -\frac{167.05}{403.60}$$
$$= -0.41 \text{ mm}$$

NOTE

Attention should be paid to the use of the correct combination of units throughout the calculation.

### 12.10.1 ALTERNATIVE: UNIVERSAL SPECTACLE MAGNIFICATION FORMULA

Instead of the solutions of example 31 shown in the previous section, equation (10.14) may be applied. The calculation will then run as follows.

*Step 1.* The tangent of the angle subtended by the object at the eye is (*Figure 12.10*)

$$\tan w = \frac{h}{-l_s + d}$$

$$= +\frac{5}{210}$$

*Step 2.* The basic retinal image size $h'$ is (equation (4.3))

$$h' = -\frac{3}{4} \times 24.53 \times \left(+\frac{5}{210}\right)$$

$$= -0.44 \text{ mm}$$

*Step 3.* Calculate the spectacle magnification from equation (10.14)

$$SM = \frac{1 - 0.01(-5.00)}{1 - 0.01(-11.00)}$$

$$= \frac{+1.05}{+1.11}$$

$$= +0.945\,9$$

*Step 4.* The retinal image size is

$$h'' = h' \times SM$$

$$= -0.44(+0.945\,9)$$

$$= -0.42 \text{ mm}$$

This value differs 0.01 mm from those calculated earlier. The difference is due to rounding-off in the course of the calculation.

### 12.10.2 ALTERNATIVE: PROXIMITY FACTOR

Bennett (1968b) shows another approach whereby the nearness of the object with respect to the eye is taken account of. The approximated proximity factor $N$ was derived as

$$N = 1 + d^2 L F_{sp} \qquad (12.13)$$

Following this approach, the total spectacle magnification for a near object consists of three factors, namely the power factor (*see* section 10.2), the shape factor (section 14.4) and the proximity factor.

Example 31 may also be solved following this approach (the shape factor being ignored).

*Step 1.* Calculate the basic retinal image size. In section 12.10.1 this was calculated to be $h' = -0.44$ mm.

*Step 2.* Calculate the spectacle magnification. For this purpose, find the far point distance:

$$F_{sp} = -6.00\text{ D} \longrightarrow \begin{aligned} f'_{sp} &= -16.67\text{ cm} \\ d &= +1.0 \quad - \\ \hline k &= -17.67 \end{aligned}$$

and by substitution into equation (10.5),

$$SM = \frac{-16.67}{-17.67}$$

$$= +0.943\,4$$

*Step 3.* The object vergence $L$ with respect to the eye, is given by (*Figure 12.10*)

$$L = \frac{1}{l_s - d}$$

$$= \frac{100}{-21}$$

$$= -4.76\text{ D}$$

*Step 4.* The proximity factor $N$ (equation 12.13)) gives

$$N = 1 + (+0.01)^2(-4.76)(-6.00)$$

$$= +1.002\,9$$

*Step 5.* The retinal image size $h''$ is given by

$$h'' = h' \times SM \times N$$

$$= -0.44(+0.943\,4)(+1.002\,9)$$

$$= -0.42\text{ mm}$$

## 12.11 The corrected eye as a telescopic system

Following the discussion in section 10.8, a further approach for the calculation of the acommodation required by a corrected ametropic eye to see a near object clearly may be developed (Obstfeld, 1976b). As shown in section 10.8, the corrected ametropic eye may be considered as a telescopic system. When a telescopic system is adjusted for infinity (*Figure 12.11*) parallel rays of light incident on the system will emerge

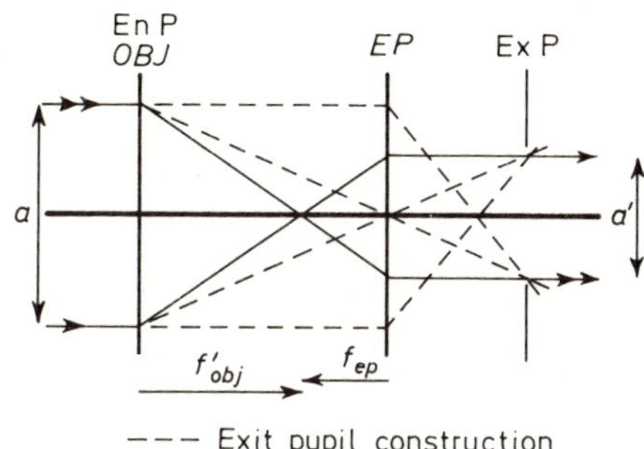


*Figure 12.11 Diagram of a simple telescopic system adjusted for infinity.* En P = *entrance pupil;* Ex P = *exit pupil; ep = eye piece; obj = objective*

again as parallel rays from the system. The entrance pupil En P of diameter *a* and the exit pupil Ex P of diameter $a'$ are conjugate planes (*see* section 20.3.3). Hence, the principal ray which enters the system at an angle *w* at the centre of the entrance pupil, emerges from the centre of the exit pupil at an angle $w'$ (*Figure 12.12*).

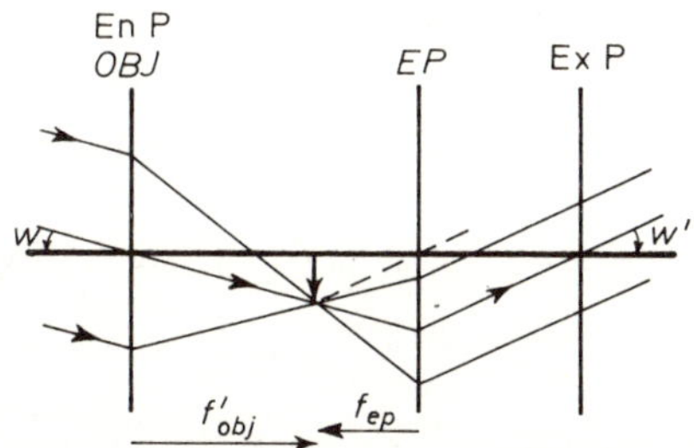


*Figure 12.12 The angular magnification of a simple telescope*

Now consider the situation where this telescope is used to observe a near object (*Figure 12.13*) of size *h* which is situated at a distance *l* from the entrance pupil, while the instrument has remained adjusted

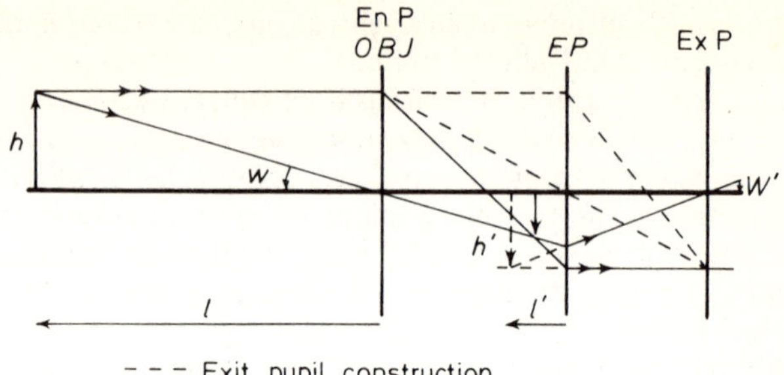


*Figure 12.13 The angular magnification of a simple telescope when used for near vision*

for infinity. Assuming an image of size $h'$ is formed at a distance $l'$ from the exit pupil, the following relationship may be developed:

$$\frac{h}{l} = w \quad \text{and} \quad \frac{h'}{l'} = w'$$

or

$$\frac{1}{l} = \frac{w}{h} \quad \text{and} \quad \frac{1}{l'} = \frac{w'}{h'}$$

By division,

$$\frac{1}{l} \Big/ \frac{1}{l'} = \frac{w}{h} \Big/ \frac{w'}{h'}$$

$$= \frac{w}{h} \times \frac{h'}{w'}$$

$$= \frac{w}{w'} \times \frac{h'}{h}$$

By inspection it will be seen (*Figure 12.11*) that

$$\frac{h'}{h} \equiv \frac{a}{a'} = M$$

and thus

$$\frac{1}{l} \Big/ \frac{1}{l'} = M^2$$

which may also be written as

$$\frac{1}{l'} = \frac{1}{l} M^2$$

or

$$L' = LM^2 \tag{12.14}$$

Note that this analysis refers to any pair of conjugate planes in the telescopic system (D.E. Gulley, 1975).

In this approach the spectacle lens has been equated with the objective and an imaginary contact lens representing the ametropia and placed in touch with an emmetropic standard reduced eye, has been equated with the eye piece. The entrance pupil of the reduced eye which is situated at the reduced surface, and the exit pupil of the telescopic system, must coincide. By definition, the entrance pupil of the telescopic system is the image formed by its objective (the spectacle lens) of its exit pupil (the pupil of the reduced eye). Hence, assuming a vertex distance of 12 mm and calculating from the exit pupil of the telescopic system towards the entrance pupil, the image vergence is

$$L' = \frac{1000}{-12}$$

$$= -83.33 \text{ D}$$

This value will only be slightly altered by the power of the average spectacle correction. Since the image vergence is so very great, the equivalent air distance will not vary a great deal with a variation of the spectacle lens power. Hence, the positions of exit and entrance pupil with respect to the reduced surface, will be virtually the same. Therefore, the distance between entrance and exit pupil may be ignored.

Equation (12.14) gives the vergence entering the reduced eye from the telescopic system. This emmetropic eye will have to accommodate when looking at a near object, at an object distance $l$. The ocular accommodation in an *emmetropic* eye is given by (*see* equation 12.3)

$$A = -L \qquad (12.15)$$

In this particular approach equation (12.14) shows that a vergence $L'$ is incident on the eye. Hence, according to equation (12.15) the eye must accommodate $-L'$ dioptres. Consequently, after substitution, equation (12.14) may be written as

$$-A = LM^2$$

or

$$A = -LM^2 \qquad (12.16)$$

As was shown in section 10.8, $M$ is the same as the spectacle magnification. Equation (12.15) may thus be written as

$$A = -L(SM)^2 \qquad (12.17)$$

not forgetting that the factor *SM* itself may be substituted by any one of its several formulae.

Applying equation (12.17) to example 29 (section 12.8), where $l = l_s - d = -10.00 - (+1.2) = -11.20$ cm, the object vergence is

$$L = \frac{100}{-11.20}$$

$$= -8.93 \text{ D}$$

while $K = -7.30$ D and $F_{sp} = -8.00$ D, one finds that the required amount of accommodation is

$$A = -(-8.93)\left(\frac{-7.30}{-8.00}\right)^2$$

$$= +7.44 \text{ D}$$

This is in reasonable agreement with the value of +7.49 D calculated earlier.

However, calculating more accurately, the power of the spectacle lens has to be taken into account. Hence, the image vergence of the exit pupil (used as object) will not be $L' = -83.33$ D as shown above, but will be $L' = -83.33 + (-7.30) = -90.63$ D, and the image distance $l' = -1.10$ cm. This represents the distance between exit and entrance pupil. Thus, the distance $l$ which was earlier given as represented by $(l_s - d)$, is in fact equal to $-10.0 - (+1.10) = -11.10$ cm. (Note that the value of 1.10 cm has changed sign since the calculation now runs in the direction opposite to that used in the calculation of the vergence $L'$.) Thus, the vergence $L$ is

$$L = \frac{100}{-11.10}$$

$$= -9.01 \text{ D}$$

Substituting this value into equation (12.17) one finds

$$A = -(-9.01)\left(\frac{-7.30}{-8.00}\right)^2$$

$$= +7.48 \text{ D}$$

which is in almost perfect agreement with the value calculated according to the step-along method.

It remains to be pointed out again that low spectacle corrections do not really produce a difference in position between entrance and exit pupil. Hence, equation (12.16) will provide an accurate value in cases of low to intermediate ametropia.

# Chapter Thirteen
# Presbyopia

## 13.1 Description

Presbyopia is a phenomenon which manifests itself in middle age. It is associated with the reduction in the amplitude of accommodation which gradually progresses from adolescence onwards. At that time it can be measured, but it is not until middle age that presbyopia becomes troublesome. Patients may then complain of burning eyes, blurring of vision, headaches etc., after prolonged near vision.

The term presbyopia (old eye) was coined by Donders (1864). Both Le Grand (1965) and Duke-Elder (1970) imply that presbyopia may be said to be present from about the age of 40 onwards. This varies from individual to individual, and the lower limit of the age of 40 may only apply to Europeans living in Europe. Studies have shown that there are racial and/or geographical variations (Coates, 1955; Rambo, 1960; Raphael, 1961; Weale, 1963). Moreover, women appear to develop presbyopia 3–5 years earlier than men (Drew, 1961; Rambo, 1953). Social considerations, such as poverty, may also influence the onset of presbyopia.

The development of presbyopia is part of the physiology of ageing. The crystalline lens is subject to sclerosis (a process of hardening) and as a result the lens' curvatures cannot be altered as much as before. Consequently, less 'accommodative power' is available, or, in other words, the amplitude of accommodation is reduced.

## 13.2 Variation of accommodation with age

Although measurements of the amplitude of accommodation of patients of different ages had been carried out before, Duane's (1912) graph is commonly used as the classical representation of the decline of the ability to accommodate as age increases (*Figure 13.1*). It should be

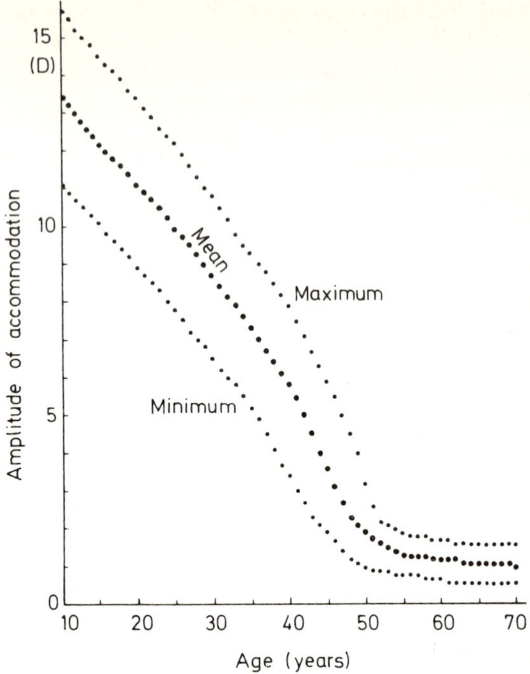


*Figure 13.1 Duane's (1912) graph showing the amplitude of monocular spectacle accommodation against age*

pointed out that his data represent monocular spectacle accommodation at a vertex distance of 14 mm.

To remember the graphical details would be rather difficult. However, since it is often useful to be able to check the amplitude of accommodation against the approximate age the following points may be used:

1. At age 13, the average amplitude of accommodation is 13 D.
2. The amplitude of accommodation decreases by about 0.25 D per year.
3. The amplitude of accommodation never seems to fall below 1 D.

## 13.3 Presbyopia and near vision

### 13.3.1 EMMETROPIA

As the amplitude of accommodation gradually decreases, the near point slowly recedes. This means that the point at which a near visual task

can be performed in comfort will also gradually move further away from the eye. As an object of given size has to be moved further away in order to be seen clearly, its retinal image size will gradually decrease and it will therefore become increasingly more difficult to recognize small detail.

What can be done to alleviate these problems for the emmetrope? By definition, an emmetrope can see clearly objects at infinity without accommodative demands being made. If it were now possible to make a near visual object appear to be situated at infinity, the problem of focusing the object on the retina would be solved.

The solution is to fit the emmetrope with a positive lens, and to advise him to place his near visual task (object) in the positive lens' first focal plane in order to fulfil the above stated conditions. If such a lens is fitted in a spectacle frame it is known as a spectacle magnifier; otherwise it is referred to as a simple magnifier, loupe, magnifying glass or simple microscope. The vertex distance of such a spectacle magnifier will be similar to that of a pair of spectacles. However, since the vertex distance has no effect on the parallel rays of light emerging from the positive lens, it could be placed at any convenient distance in front of an emmetropic eye as long as the object remained in the first focal plane of the lens.

What is the effect of this solution on the retinal image size? Two approaches may be followed, namely:

1. Compare the retinal image size produced by the spectacle magnifier eye system, with that of the retinal image size produced by the eye on its own; or
2. Compare the angular subtense of the object's image at the spectacle magnifier, with that of the object placed at the 'least distance of distinct vision' from the eye.

Following the first approach, it must be pointed out that only the angles subtended by the chief ray at the principal point P of the eye, need to be compared (*see* section 10.2). Thus, if an object of given size is viewed from a distance $l$, it will subtend an angle $w_o$ where (*Figure 13.2*)

$$w_o = \frac{h}{-l}$$

The same object will subtend an angle $w_s$ when seen through the spectacle magnifier where (*Figure 13.3*)

$$w_s = \frac{h}{f'}$$

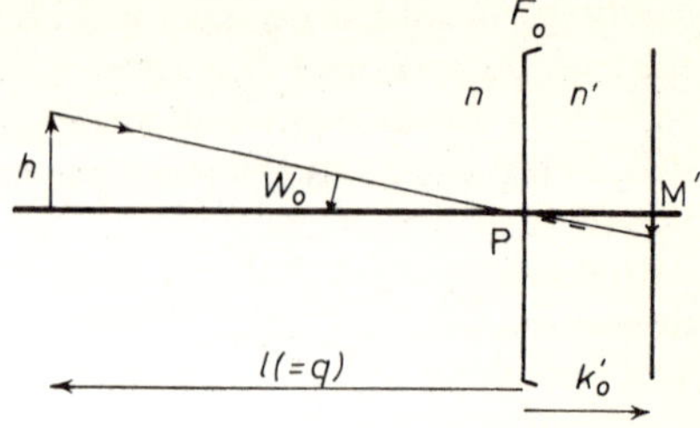


*Figure 13.2 The angle $w_o$ subtended by a near object at the eye*

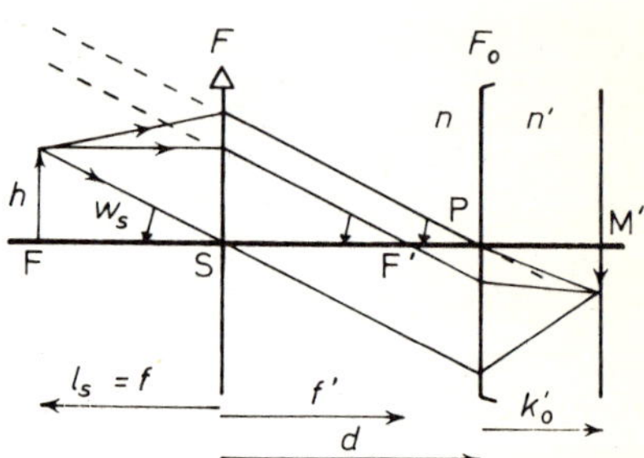


*Figure 13.3 The angle $w_s$ subtended by a near object viewed through a spectacle magnifier, at the eye*

The magnification is then given by

$$
\begin{aligned}
M &= \frac{\tan w_s}{\tan w_o} \\
&= \frac{h}{f'} \bigg/ \frac{h}{-l} \\
&= \frac{h}{f'} \times \frac{-l}{h} \\
&= -\frac{l}{f'} \\
&= -\frac{l}{f'} \qquad (13.1)
\end{aligned}
$$

Thus, the magnification depends on the object distance $l$ and the power of the spectacle magnifier.

The second approach assumes the use of a standard 'least distance of distinct vision', denoted $q$. This distance is usually taken to be 25 cm, measured from the eye, and is chosen because it is said to be a distance at which an object can be viewed comfortably and without undue effort. (Recently, it has been suggested that it would be more realistic to assume a distance of $\frac{1}{3}$m.) The magnification calculated on this basis

is referred to as apparent magnification, magnifying power or loupe magnification. It is calculated thus:
The object is placed at the least distance of distinct vision. Hence, $l = q$. The angular subtense of the object at this distance is (*Figure 13.2*)

$$\tan w_o = \frac{h}{-q} = \frac{h}{-(-25)}$$

while the angular subtense of this object seen through the spectacle magnifier is given by (*Figure 13.3*)

$$\tan w_s = \frac{h}{-f}$$

The magnifying power of the spectacle magnifier is then

$$M = \frac{\tan w_s}{\tan w_o} = \frac{h}{-f} \bigg/ \frac{h}{+25} = -\frac{25}{f} = \frac{25}{f'}$$

where $f'$ must be expressed in centimetres. This formula may be transformed into the more frequently used formula employing dioptres rather than a measure of length, thus

$$M = 25 \bigg/ \frac{100}{F} = 25 \times \frac{F}{100} = \frac{F}{4} \qquad (13.2)$$

EXAMPLE 32

Calculate the angular magnification as well as the magnifying power of a system consisting of an emmetropic eye and a spectacle magnifier. The object under observation is situated 20 cm in front of the eye. The magnifier has a power of +10.00 D, or a second focal length of +10.00 cm

*Solution 32*
Angular magnification (equation (13.1)):

$$M = \frac{-20.00}{-(+10.00)}$$
$$= +2.000\ 0$$

Magnifying power (equation (13.2)):

$$M = \frac{+10.00}{4}$$
$$= +2.500\ 0$$

EXAMPLE 33
Calculate the angular magnification and the magnifying power of an emmetropic eye – spectacle magnifier system. The object distance is −5.00 cm, and the magnifier has a power of +10.00 D ($f' = +10.00$ cm).

*Solution 33*
Angular magnification (equation (13.1)):

$$M = \frac{-5.00}{-(+10.00)}$$
$$= +0.500\ 0$$

Magnifying power (equation (13.2)):

$$M = \frac{+10.00}{4}$$
$$= +2.500\ 0$$

NOTE
In both example 32 and 33 the eye would need to accommodate if it were to observe the object without the aid of the magnifier. Whether or not it can accommodate is of no consequence with respect to the magnification, since it is the basic retinal image size (determined by the chief ray) which is used for the purpose of comparison (*see* section 10.2).

NOTE
The angular magnification found in the solution of example 33, is smaller than unity. This indicates that the basic retinal image in the

unaided eye is larger than that of the eye when aided by the magnifier! This is due to the angular subtense of the object at the eye when held very close to the eye. Uncorrected ametropes, both presbyopic and non-presbyopic, as well as emmetropic presbyopes sometimes use this phenomenon. They usually discover it by accident. The magnitude of the basic retinal image then outweighs the effect of blurring in the ability to interpret the retinal image.

Now, if an eye can accommodate and the object is placed at a distance $l$ which is smaller than the first focal length $f$ of the spectacle magnifier, the situation is changed altogether. The position of the apparent object (that is the image at B$'$ of size $h'$) may be constructed as shown in *Figure 13.4.*

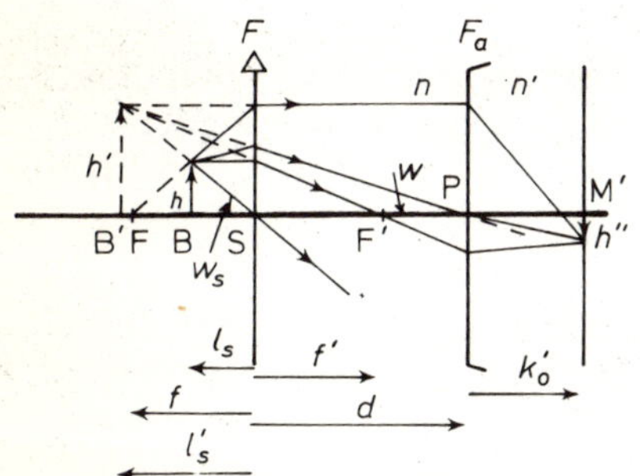


*Figure 13.4 The angle w subtended by a near object viewed through a magnifier, at an accommodated eye*

Assuming that the amplitude of accommodation of this eye is sufficient to image the object on the retina, what magnification is obtained under these circumstances? If the first approach (*see above*) is followed once again, the following reasoning applies:

If the spectacle magnifier were not used, the angle subtended by the object at the reduced surface would be described by (*Figure 10.8*)

$$\tan w_o = \frac{h}{-l_s + d}$$

$$= \frac{h/(-1 + dL_s)}{L_s}$$

$$= \frac{hL_s}{-1 + dL_s}$$

With the spectacle magnifier in place the angle subtended by the apparent object at the reduced surface may be calculated thus:

The apparent object at B$'$ subtends an angle $w$ at P where (*Figure 13.4*)

$$\begin{aligned}\tan w &= \frac{h'}{-l'_s + d} \\ &= \frac{h'/(-1 + dL'_s)}{L'_s} \\ &= \frac{h'L'_s}{-1 + dL'_s}\end{aligned}$$

According to equation (1.15), $h'$ may be represented by

$$h' = \frac{L_s}{L'_s} h$$

Furthermore, in accordance with the conjugate foci formula (equation (1.12)), the image vergence $L'_s$ is given by

$$L'_s = L_s + F$$

so that, by substitution, one may write

$$\tan w = \frac{\dfrac{L_s}{L_s + F} \times h(L_s + F)}{-1 + d(L_s + F)}$$

which can be simplified to

$$\tan w = \frac{L_s h}{-1 + d(L_s + F)}$$

The magnification of this system is given by

$$\begin{aligned}M &= \frac{\tan w}{\tan w_o} \\ &= \frac{hL_s}{-1 + d(L_s + F)} \bigg/ \frac{hL_s}{-1 + dL_s} \\ &= \frac{hL_s}{-1 + d(L_s + F)} \times \frac{-1 + dL_s}{hL_s} \\ &\simeq \frac{-1 + dL_s}{-1 + d(L_s + F)} \\ &= \frac{1 - dL_s}{1 - d(L_s + F)} \qquad (13.3)\end{aligned}$$

which is another way of writing the Universal Spectacle Magnification formula (equation (10.14)).

Applying this formula, the following values may be used as an example:

EXAMPLE 34

Calculate the magnification of the retinal image for an object placed −12.50 cm ($L_s$ = −8.00 D) in front of a lens of +5.00 D. The latter is held 15 mm in front of the eye.

*Solution 34*

Substituting the data into equation (13.3), the magnification is found to be

$$M = \frac{1 - 0.015(-8.00)}{1 - 0.015(-8.00 + 5.00)}$$

$$= \frac{1 + 0.12}{1 + 0.045}$$

$$= \frac{+1.12}{+1.045}$$

$$= +1.071\,8$$

However, readers might find it hardly worthwhile to memorize another formula when the answer can easily be found from first principles, thus (*Figure 10.8*)

$$\tan w_o = \frac{h}{-l_s + d}$$

$$= \frac{h}{+14.00}$$

Similarly (*Figure 13.4*)

$$\tan w = \frac{h'}{-l'_s + d}$$

The distance $l'_s$ may be calculated using, for instance, the step-along method:

$$\begin{array}{lll} L_s = -8.00\ \text{D} & \longleftarrow & l_s = -12.50\ \text{cm} \\ F = +5.00 & + & \\ \hline L'_s = -3.00 & \longrightarrow & l'_s = -33.333 \end{array}$$

The magnification is then given by

$$m = \frac{L_s}{L'_s} = \frac{-8.00}{-3.00} = +2.666\,7$$

Since $h' = mh$, one may write

$$\tan w = \frac{+2.666\,7}{-(-33.33) + 1.5} = \frac{+2.666\,7}{+34.83}$$

Now

$$M = \frac{\tan w}{\tan w_o} = \frac{(+2.666\,7)h}{+34.83} \bigg/ \frac{h}{+14.00} = \frac{(+2.666\,7)h}{+34.83} \times \frac{+14.00}{h} = \frac{+37.33}{+34.83} = +1.071\,8$$

This concludes the calculations for the first approach.

The second approach (*see above*) is based on the concept of magnifying power and employs the 'least distance of distinct vision'. This means that the apparent object as seen through the spectacle magnifier, is compared with the object placed at the 'least distance of distinct vision'. In order to see an object placed within the first focal length of the spectacle magnifier, an amount of spectacle accommodation $A_s$ needs to be exerted. This is measured in the spectacle plane, that is, in the plane of the spectacle magnifier. For this purpose one may imagine that the dioptric power resulting from the accommodation is added to the power of the spectacle magnifier. Hence, the magnifying power of

a spectacle magnifier used by an accommodating emmetropic eye, is given by

$$M = \frac{F + A_s}{4} \tag{13.4}$$

However, it might be more convenient to remember the formula as

$$M = \frac{F - L_s}{4} \tag{13.5}$$

where the object vergence is substituted for the spectacle accommodation (equation (12.11)).

Applying equation (13.5) to the data of example 34 one finds

$$\begin{aligned} M &= \frac{+5.00 - (-3.00)}{4} \\ &= +2.000\,0 \end{aligned}$$

The reader will note the very different results obtained following each approach. However, these results are not comparable since they are based on two different concepts.

Returning to the theme of this section, namely accommodation and presbyopia, consider the amount of spectacle and of ocular accommodation required under the circumstances set out in example 34 (*Figure 13.4*).

The spectacle accommodation may be calculated from equation (12.11):

$$\begin{aligned} A_s &= -(-8.00) \\ &= +8.00 \text{ D} \end{aligned}$$

The ocular accommodation may be found using equation (12.12) or the step-along method. The calculation is left to the reader; the outcome is +2.87 D.

However, if this emmetropic eye had not been fitted with a spectacle magnifier it would have had to accommodate (*see Figure 13.4*)

$$\begin{aligned} A &= -\frac{1}{l_s - d} \\ &= -\frac{100}{-12.50 - (+1.5)} \\ &= -\frac{100}{-14.00} \\ &= +7.14 \text{ D} \end{aligned}$$

which is over 4 D more! The magnification of the retinal image would have been very different, too, as is shown below. Since this is an emmetropic eye, the accommodation also indicates the vergence incident on the eye too, thus

$$L = -A$$
$$= -(+7.14)\ \mathrm{D}$$

so that, according to equation (12.10), the magnification is given by

$$m = \frac{-7.14}{+60.00}$$
$$= -0.119\,0$$

This value should be compared with that found for the retinal image magnification by the spectacle magnifier and the eye's optical system, as follows.

It has been shown that the magnification due to the spectacle magnifier is $M = +1.071\,8$. The magnification by the eye's optical system is given by equation (12.10).

$$m = \frac{-2.87}{+60.00}$$
$$= -0.047\,8$$

Thus, the magnification of the system is

$$Mm = (+1.071\,8)(-0.047\,8)$$
$$= -0.051\,2$$

The collection of data calculated above may be recapitulated as follows:

| magnifying power | magnification aided eye | magnification unaided eye |
|---|---|---|
| $M = +2.000\,0$ | $Mm = -0.051\,2$ | $m = -0.119\,0$ |
| $A = +2.87$ D | $A = +2.87$ D | $A = +7.14$ D |

The reader will appreciate that these are only some of the factors that will decide whether or not to employ a spectacle magnifier. Considerations of a practical nature as well as cosmetic aspects may have an over-riding effect in practice.

### 13.3.2 UNCORRECTED MYOPIA

In myopia both the far point and the near point are situated in front of the eye. As the amplitude of accommodation decreases gradually, the near point will be found to move slowly away from the eye towards the far point. Since the amplitude of accommodation never seems to fall below 1 D, the near point distance will always remain shorter than the far point distance. Consider the following example.

EXAMPLE 35
A myope of 1.00 D ocular correction could accommodate 8 D at the age of 36, 4 D at the age of 44 and 1 D at the age of 52 years. Calculate the far point and near point distances at each age.

*Solution 35*
The far point distance depends only on the ocular correction. Therefore, the far point distance remains the same during these years, namely (equation (5.7))

$$\begin{aligned} k &= \frac{100}{-1.00} \\ &= -100.00 \text{ cm} \end{aligned}$$

NOTE
In this example it has been assumed that the ocular correction does not change in the course of some 15 years. In practice small variations would probably have been encountered.

The near point position may be calculated on the basis of equations (12.5) and (12.4). Thus, at age 36, the near point vergence is found to be

$$\begin{aligned} B &= -1.00 - (+8.00) \\ &= -9.00 \text{ D} \end{aligned}$$

and the near point distance is

$$\begin{aligned} b &= \frac{100}{-9.00} \\ &= -11.11 \text{ cm} \end{aligned}$$

At age 44:

$$B = -1.00 - (+4.00)$$
$$= -5.00 \text{ D}$$

and so

$$b = -20.00 \text{ cm}$$

At age 52:

$$B = -1.00 - (+1.00)$$
$$= -2.00 \text{ D}$$

so that

$$b = -50.00 \text{ cm}$$

The withdrawal of the near point may be shown graphically as in *Figure 13.5*. Clear vision is only possible when the object is situated

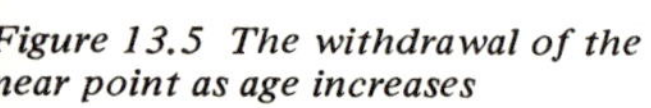

*Figure 13.5 The withdrawal of the near point as age increases*

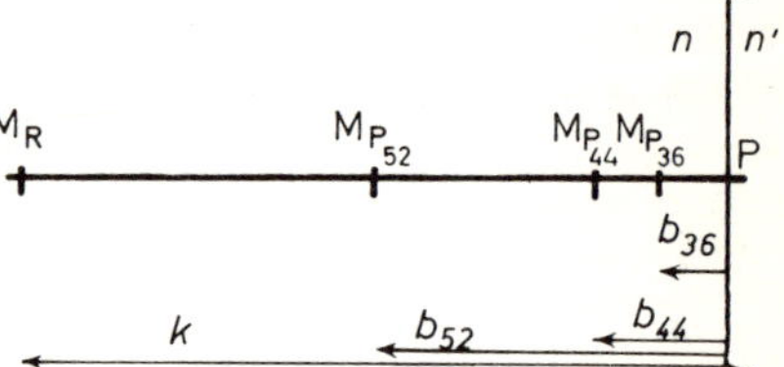


between the far point and the near point. As the near point approaches the far point, the 'range of accommodation' (that is, the distance between far point and near point, measured in units of length) becomes smaller. In this example the range is 89 cm at age 36, it is still 80 cm at age 44, but has been reduced to 50 cm at age 52. These 50 cm are situated almost beyond a comfortable arm's length distance. This means that the patient will not be able to read or write for any length of time without experiencing symptoms of discomfort.

The above discussed case applied to low myopia. For an intermediate degree of myopia the situation with respect to near vision may be more favourable. If a patient has an ocular correction of between 2.5 and 5 D he is not likely to experience much difficulty with near vision. His far point would be situated somewhere between −25 and −40 cm. This is

a comfortable near vision distance for most people. These uncorrected myopes will not need a near vision correction; in fact, they often remove their distance correction (if worn) when doing close work.

However, eyes having a higher degree of myopia would not be able to see clearly objects within the normal range of reading distances (say, from −25 to −40 cm). This would not cause problems to people who have only one functioning eye. In binocular vision, however, both eyes' visual axes will have to be directed at the far point where the object may be assumed to be placed. This will require a large effort by the extra-ocular muscles, in particular by the medial recti. Since the far point (as well as the near point) will be at a distance of less than −25 cm from the eye, the eyes will have to converge an inordinate amount. This does not normally cause problems when required for a short while only, but it is unlikely to be maintained without discomfort for any length of time. When the eyes' visual axes come to meet at a point situated beyond the far point, the object (situated at the far point) will be seen double (double vision or diplopia) and the retinal images will be blurred.

### 13.3.3 CORRECTED MYOPIA

When any eye is corrected, it means that a distant object is imaged on the retina of that eye with the aid of the correction. If the eye can accommodate, objects situated nearer by than infinity can also be sharply imaged on the retina of that eye.

As has been explained, the amplitude of accommodation decreases slowly as the patient's age increases. Consequently, in the corrected ametropic eye, the apparent near point moves further and further away from the eye. In this context the effect of the spectacle correction and vertex distance must be considered, as in the following example.

#### EXAMPLE 36

Spectacle correction −8.00 D. Vertex distance 15 mm. Amplitude of accommodation 9.00 D, but some year later only 3.00 D. Calculate the apparent near point positions (*Figure 13.6*).

*Solution 36*

*Step 1.* Calculate the ocular correction. Using the step-along method:

$$F_{sp} = -8.00\text{ D} \longrightarrow f'_{sp} = -12.50\text{ cm}$$

$$d = +1.5 \quad -$$

$$K = -7.14 \longleftarrow k = -14.00$$

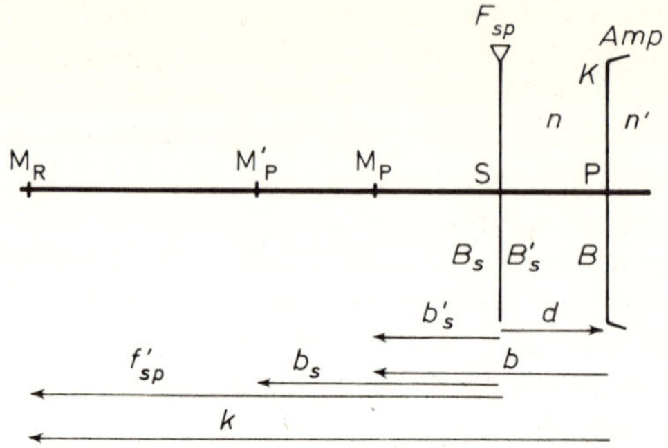


*Figure 13.6 Example 36; far point* $M_R$*, near point* $M_P$ *and apparent near point* $M'_P$ *in myopia*

*Step 2.* If the ocular accommodation of 9.00 D is exerted, the vergence incident on the eye that will be focussed on the retina (the near point vergence), is (equation *12.4*)

$$\begin{aligned} B &= K - Amp \\ &= -7.14 - (+9.00) \\ &= -16.14 \text{ D} \end{aligned}$$

*Step 3.* The image vergence $B'_s$ emerging from the spectacle lens that will produce this vergence at the eye, is:

$$B = -16.14 \text{ D} \longrightarrow \begin{aligned} b &= -6.20 \text{ cm} \\ d &= +1.5 \quad + \\ \hline b'_s &= -4.70 \end{aligned}$$

$$B'_s = -21.28 \longleftarrow b'_s = -4.70$$

*Step 4.* The vergence $B_s$ incident on the spectacle lens to give this vergence $B'_s$ after refraction by this lens, may be calculated from the conjugate foci formula:

$$\begin{aligned} B'_s &= -21.28 \text{ D} \\ F_{sp} &= -8.00 \quad - \\ \hline B_s &= -13.28 \end{aligned}$$

*Step 5.* The apparent near point $M'_P$ is, therefore, situated at a distance $b_s$ of (equation 12.4)

$$\begin{aligned} b_s &= \frac{100}{-13.28} \\ &= -7.53 \text{ cm} \end{aligned}$$

NOTE
The near point and apparent near point are conjugate through the spectacle lens.

In order to find out how this has changed some time later when the amplitude of accommodation is only 3.00 D, the calculation from Step 2 onwards needs to be repeated, taking into account the reduced amplitude. The result of this calculation shows that the apparent near point is now situated at $b_s = -25.25$ cm.

What are the implications of these figures? When the eye could still accommodate 9.00 D, its apparent near point was at about $b_s - d = -7.5 - (+1.5) = -9$ cm from the eye. That would be comparable to some 11 D of accommodation in an emmetropic eye, or it would appear as if this corrected myopic eye had 2 D more accommodation than an emmetrope. When the amplitude had decreased to 3.00 D, the apparent near point was at about $b_s - d = -25 - (+1.5) = -26.5$ cm from the eye. This would be comparable with 3.75 D of accommodation in an emmetropic eye. Thus, the effect of accommodation in the corrected myopic eye is enhanced by its spectacle correction. This means that a corrected myope can manage without 'reading glasses' longer than an emmetrope who has the same amplitude of accommodation. However, when the spectacle correction is replaced by a contact lens correction, the myope's advantage disappears since this change has made him virtually emmetropic without having affected the presbyopia.

### 13.3.4 UNCORRECTED HYPEROPIA

The far point of the hyperopic eye is situated 'beyond' infinity. Theoretically, the far point lies behind the eye. If the amplitude of accommodation of such an eye is equal to its degree of hyperopia, its near point will lie at infinity. However, if the amplitude of accommodation is larger than the degree of hyperopia, accommodation becomes of practical use for near vision as the near point will now be found between infinity and the eye. A visual task can then be performed at such a near point distance since the object will be imaged clearly on the retina.

As time goes by, the amplitude of accommodation will slowly decrease and so the near point distance will slowly increase. Here, the degree of hyperopia is of great importance: if the eye is highly hyperopic, it is unlikely that accommodation can be maintained for a long period of time without the patient being subject to symptoms of strain. This applies even more as the amplitude becomes smaller.

A further problem encountered by patients who have binocular vision, is that there is a certain relationship between accommodation and convergence (the movement made by a pair of eyes in order to direct the visual axes to intersect at a near object). When an uncorrected hyperope of, say, 7 D ocular correction wants to see a distant object clearly, the eyes will have to accommodate 7 D but should not converge! Compare this with an emmetrope who accommodates 7 D. At that moment the retina will be conjugate with a point some 14 cm in front of the eye *and* the eyes have to converge so that their visual axes will intersect at that point.

On the other hand, if the degree of hyperopia is low, say 2 D, and the eyes can accommodate 7 D, the retina will be conjugate with a point 20 cm in front of the eye. The eyes' visual axes will have to be converged to this point too. In this case the discrepancy between amount accommodated and amount converged is not so great, and it is possible that this patient will be able to maintain binocular single vision for a long time.

However, low and intermediate hyperopes become subject to symptoms of 'pre-presbyopia' (similar to presbyopic symptoms, but encountered at a younger age than that at which presbyopia is usually experienced). In most cases these problems can be alleviated by correcting the hyperopia, as a first measure. When presbyopia proper sets in, further help in the form of a near vision addition (*see below*) needs to be provided.

### 13.3.5 CORRECTED HYPEROPIA

The artificial far point of a corrected hyperope will be situated at infinity. Any accommodation available will bring the point in space which is conjugate with the retina, closer to the eye. The effect of the spectacle correction and the vertex distance must be taken into account. Consider the following example.

EXAMPLE 37

Spectacle correction +8.00 D. Vertex distance 15 mm. Amplitude of accommodation 9.00 D, decreasing to 3.00 D after several years. Calculate the apparent near point distances (*Figure 13.7*).

*Solution 37*

*Step 1.* Calculate the ocular correction:

$$F_{sp} = +8.00 \longrightarrow f'_{sp} = +12.50 \text{ cm}$$

$$d = +1.5 \quad -$$

$$K = +9.09 \longleftarrow k = +11.00$$

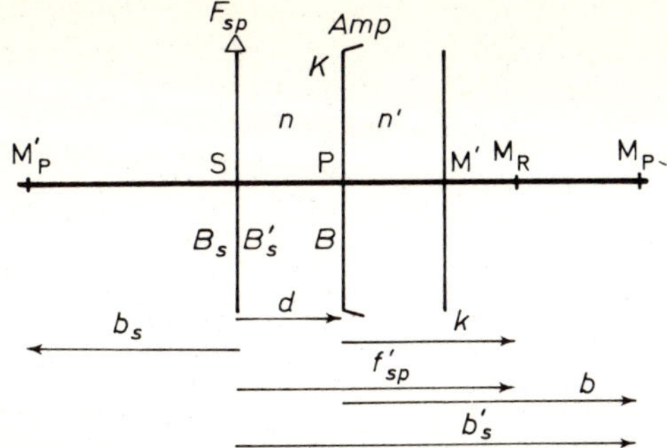


*Figure 13.7 Example 37; far point, near point and apparent near point in hyperopia*

*Step 2.* The near point vergence may be calculated from equation (12.5)

$$\begin{aligned} B &= K - Amp \\ &= +9.09 - (+9.00) \\ &= +0.09 \text{ D} \end{aligned}$$

*Step 3.* The vergence leaving the spectacle lens to produce the near point vergence at the eye, will have to be

$$B = +0.09 \text{ D} \longrightarrow b = +1111.11 \text{ cm}$$

$$d = +1.5 \quad +$$

$$B'_s = +0.09 \longleftarrow b'_s = +1112.61$$

NOTE

Since the vergence $B$ is so small, and the air distance, therefore, so very long, the vertex distance does not have much effect.

*Step 4.* Hence, the vergence incident on the lens must be (equation (1.12))

$$\begin{aligned} B_s &= +0.09 - (+8.00) \\ &= -7.91 \text{ D} \end{aligned}$$

*Step 5.* The apparent near point is situated at a distance (equation (12.4))

$$\begin{aligned} b_s &= \frac{100}{-7.91} \\ &= -12.64 \text{ cm} \end{aligned}$$

from the spectacle plane.

Some years later the situation has altered because of the decreasing amplitude of accommodation. The apparent near point is then found to be situated at a distance of −41.32 cm from the spectacle plane.

The effect of 9 D of accommodation and a spectacle lens of +8.00 D at a vertex distance of 15 mm is an apparent near point distance from the eye of $b_s - d = -12.5 - (+1.5) = -14$ cm (approx.). This distance would be comparable with an emmetrope accommodating 7 D. The effect of this spectacle correction, therefore, represents a 'loss' of 2 D of accommodative power as compared with the accommodation exerted.

Some years later the apparent near point is at about $b_s - d = -41.5 - (+1.5) = -43$ cm from the eye. This would be comparable with an emmetrope accommodating $2\frac{1}{3}$ D, hence, a 'loss' of accommodational effect of about $\frac{2}{3}$ D.

In short, the effect of accommodation in the corrected hyperopic eye is decreased by its spectacle correction. This means that a corrected hyperope will require a near vision addition (*see below*) earlier in life than an emmetrope who has the same amplitude of accommodation. However, when the spectacle correction is replaced by a contact lens correction the hyperope's disadvantage is virtually nullified since this change has made the eye 'emmetropic' for all practical purposes.

## 13.4 Theoretical calculation of the near vision addition

As has been indicated in the previous sections the whole amplitude of accommodation cannot be used for a prolonged period of time without the patient experiencing some discomfort. In order to alleviate these symptoms only part of the available accommodation should be used continuously. Any deficiency in the amount of accommodation available for prolonged use and the amount required in view of the working distance, is then made up by means of a 'near vision addition'. This is usually referred to as the 'reading addition', 'reading add' or, simply, 'the add'. This is the amount of positive focal power added to the distance prescription (if any) for near vision purposes in presbyopia. In accordance with BS 3521 : 1962, it will be abbreviated to '*Add*'.

Since the addition modifies the power of the spectacle lens (or part of that lens depending on the type of spectacle lens employed for this purpose), the accommodation required as well as the accommodation available for prolonged use must be measured in the spectacle plane.

The theoretical calculation of the addition is based on the clinical observation that about two-thirds of the available spectacle accommodation may be exerted continuously without discomfort being experienced. The amount of accommodation required will depend on the

working distance or near vision distance $l_s$ from the spectacle plane. Thus, the addition may be calculated from the formula

$$Add = -(L_s + \frac{2}{3}Amp) \qquad (13.6)$$

EXAMPLE 38

A patient's normal working distance is −25 cm from his spectacle plane. The amplitude of accommodation, measured in the spectacle plane, is 4.50 D. Calculate the theoretical near vision addition.

*Solution 38*

Substituting the data provided into equation (13.6),

$$\begin{aligned} Add &= -[-4.00 + \frac{2}{3}(+4.50)] \\ &= +1.00 \text{ D} \end{aligned}$$

NOTE

This patient will be able to see comfortably and clearly objects at less than 25 cm from his spectacle plane for short periods of time. When relaxing his accommodation he may bring objects situated at any distance up to −1 m (due to the 1 D add) in focus on his retina. Hence, there is a region, theoretically from −1 m to infinity, in which objects cannot be imaged clearly. This may prove a great problem under certain circumstances (reading labels on articles on shelves, operating machines etc.). For such purposes an 'intermediate add' may be prescribed where the add is equal to, say, half the addition. As the addition increases in power, the furthest distance that can be seen clearly through this (part of the) lens, approaches the spectacle plane. Hence, intermediate additions will also be required for such occupations and hobbies as playing the piano, reading music, playing bridge and painting. Alternatively, the working distance associated with the particular occupation is measured, and the addition calculated on that basis.

NOTE

Equation (13.6) provides only an indication of the addition required. A clinically determined addition may well be different.

## 13.5 Depth of field

The depth of field may be defined as the distance over which an object may be moved along the axis of an optical system without causing its

image to move away from the correct image plane by more than a given tolerable amount. In this context it is of interest to consider the definition of the depth of field's counterpart, namely the depth of focus. This may be defined as the distance on either side of the best focus of an optical system, over which the sharpness of the image does not fall below a given tolerable amount. It is usually specified by the maximum diameter of the blur circle. However, in visual optics both depth of field and depth of focus are usually specified in dioptres.

What is the importance of the depth of field? If an optical system did not have any depth of field, any object not placed in its object plane would be imaged out of focus in the system's image plane. With respect to the eye, it would mean that any object not situated at its far point, its near point or the area between these two points, would appear to be blurred. Fortunately, this is not so. Investigations (Leinhos, 1959) have shown that young people's eyes have a depth of field of about $\frac{1}{3}$ D. This increases to $\frac{3}{4}$ D in old age. One may, therefore, assume as an average value a depth of field of 0.5 D in presbyopes. *See also* Schober (1970).

The depth of field depends on the pupil diameter, visual acuity, contrast and object distance (*see also* Campbell, 1957, 1959; Rantzsch, 1964).

The depth of field of non-presbyopes is of lesser interest since their amplitude of accommodation will usually be sufficient to deal with intolerable blurring of the retinal image.

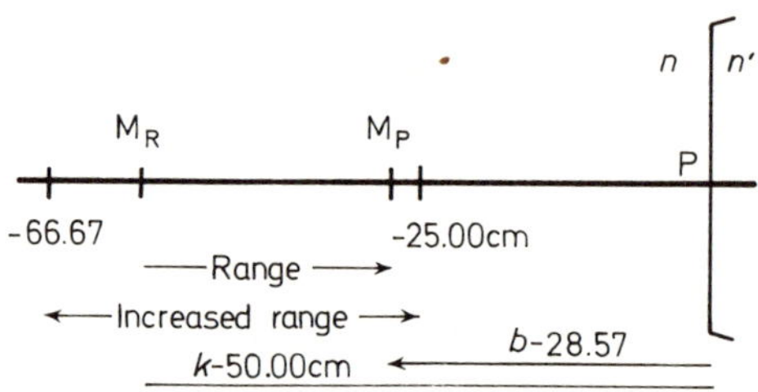


*Figure 13.8 The effect of the depth of field on the range of accommodation*

Now consider the effect of the depth of field in an uncorrected myopic eye of 2.00 D ocular correction, with an amplitude of accommodation of 1.50 D (*Figure 13.8*). In this case the far point distance is (equation (5.7))

$$k = \frac{100}{-2.00}$$

$$= -50.00 \text{ cm}$$

However, because of the depth of field (previously assumed to be 0.50 D), the vergence arriving at the eye may be modified by moving the object from the exact far point. If the object is moved away from the eye, the object distance will be increased and the vergence arriving at the eye will thus be decreased. The vergence may be reduced to $-2.00 + 0.50 = -1.50$ D. The object is then placed $-66.67$ cm from the eye. Any reduction in the far point distance by a distance equivalent to less than the amplitude of accommodation, will be taken care of by accommodating. The near point vergence may be calculated from equation (12.5):

$$\begin{aligned} B &= -2.00 - (+1.50) \\ &= -3.50 \text{ D} \end{aligned}$$

Hence, the near point distance is (equation (12.4)):

$$\begin{aligned} b &= \frac{100}{-3.50} \\ &= -28.57 \text{ cm} \end{aligned}$$

Now, because of the depth of field, the near point vergence may be increased by 0.50 D so that a vergence of $-3.50 - 0.50 = -4.00$ D is incident on that eye. (To decrease it would require a relaxation of accommodation which does not pose a problem.) This vergence is associated with an object distance of $-25$ cm.

Thus, the range of accommodation is equal to

$$\begin{aligned} |k| - |b| &= 50 - 28.57 \\ &= 21.43 \text{ cm} \end{aligned}$$

but when the depth of field is taken into consideration, the range is increased to

$$66.67 - 25 = 41.67 \text{ cm}$$

In this case the modified range, that is, the range modified by the depth of field, is virtually double the original range of accommodation.

Now, consider the following example of a high myope.

EXAMPLE 39

A myope of 8.00 D ocular correction has an amplitude of accommodation of 2.00 D and a depth of field of 0.50 D. Calculate the range of

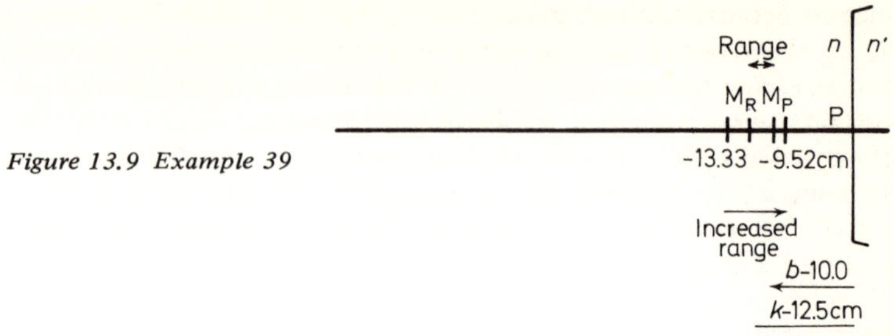


*Figure 13.9 Example 39*

accommodation. To what extent is it modified by the depth of field? (*Figure 13.9*).

*Solution 39*

*Step 1.* Calculate the far point distance (equation (5.7)):

$$k = \frac{100}{-8.00}$$

$$= -12.50 \text{ cm}$$

*Step 2.* Calculate the near point distance (equations (12.5) and (12.4)):

$$B = -8.00 - (+2.00)$$

$$= -10.00 \text{ D}$$

and

$$b = -10.00 \text{ cm}$$

*Step 3.* The range of accommodation is given by

$$|k| - |b| = 12.50 - 10.00$$

$$= 2.50 \text{ cm}$$

*Step 4.* The depth of field modifies the ocular correction from −8.00 D to −7.50 D. This is associated with a distance of −13.33 cm.
*Step 5.* The depth of field modifies the near point vergence from −10.00 D to −10.50 D which is associated with a distance of −9.52 cm.
*Step 6.* The depth of field increases the range of accommodation to

$$13.33 - 9.52 = 3.81 \text{ cm}$$

Although the effect of the depth of field is only a slight increase in the range of accommodation (namely by 1.3 cm), nevertheless it represents an increase of ~50%!

In uncorrected hyperopia the depth of field is of no practical value with respect to the far point distance since this is situated beyond infinity. If the amplitude of accommodation is smaller than the hyperopia, it is of no value with respect to the near point either. However, if the uncorrected eye can overcome its hyperopia by accommodating and, after doing so, has a few dioptres in reserve, the depth of field will be very useful as it will increase the range of accommodation. Consider the following example.

EXAMPLE 40

Ocular correction +5.00 D. Amplitude of accommodation 8.00 D.

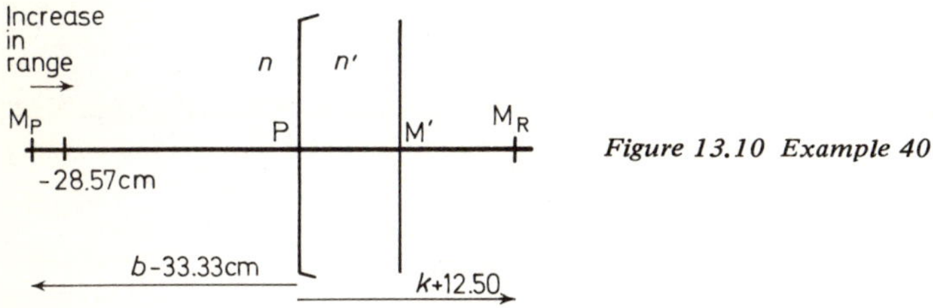


*Figure 13.10 Example 40*

Calculate the near point distance. What effect has 0.50 D depth of field? (*Figure 13.10*).

*Solution 40*

*Step 1.* The near point vergence is given by equation (12.5):

$$
\begin{aligned}
B &= +5.00 - (+8.00) \\
&= -3.00 \text{ D}
\end{aligned}
$$

and the near point distance is (equation (12.4)):

$$b = -33.33 \text{ cm}$$

*Step 2.* In view of the depth of field the near point vergence may be modified to −3.50 D which is associated with a distance of −28.57 cm. This shows that the nearest distance at which an object can be seen sufficiently clearly is about 5 cm nearer to the eye than the near point position.

In cases of corrected ametropia one ought to consider the depth of field with respect to the vergence incident on the reduced surface. However, since the depth of field is no more than an approximation of the actual situation it will make little difference in clinical practice

when, for the purpose of a simple calculation, the dioptric value of the depth of field is applied to the vergence reaching the spectacle plane rather than the eye itself.

With respect to hyperopia, the range of accommodation poses a problem: the distance between far point and near point cannot really be given in terms of units of length, since the far point is virtual. Assuming that there is a sufficient amount of accommodation available for the near point to be situated in front of the eye, it is best to indicate the range of accommodation by saying that it runs from 'infinity to − ... cm' (the latter being the position of the near point).

## 13.6 'Simultaneous equations'

A favourite type of question requires the student to find two equations which enable him to calculate the required information. When reading the question the uninitiated reader may get the impression that the required information has nothing to do with the details provided. However, the details that have been omitted should be available from memory to those who have studied this chapter. Two examples follow here.

EXAMPLE 41

An eye has originally a far point distance of −1 m and a near point distance of −10 cm. Several years later the far point distance has increased by 50% and the near point distance is now a third of the far point distance. Estimate the increase in the age of the eye.

*Solution 41*

Original situation:

From the far point distance and the near point distance, the ocular correction and near point vergence may be calculated, and hence the original amplitude of accommodation (equation (12.5)). Thus:

$$\begin{aligned} Amp &= K - B \\ &= \frac{100}{-100.00} - \frac{100}{-10.00} \\ &= -1.00 - (-10.00) \\ &= +9.00 \text{ D} \end{aligned}$$

New situation:

The far point distance has increased to −1.50 m and the near point

distance to $\frac{1}{3}(-1.50) = -0.50$ m. Hence, the amplitude of accommodation is (equation (12.5))

$$Amp = \frac{100}{-150.00} - \frac{100}{-50.00}$$
$$= -0.67 - (-2.00)$$
$$= +1.33 \text{ D}$$

Hence, over the years the amplitude has decreased by $+9.00 - (+1.33) = +7.67$ D. Since it may be assumed that the amplitude of accommodation decreases 0.25 D per year, the decrease in the amplitude of this eye took place in the course of $+7.67/+0.25 = \sim 30$ years.

It would also be possible to phrase the same basic question in a very different way, namely:

Originally, the far point distance of an eye is −1 m, and its near point vergence is 10 times the ocular correction. Several years later, the ocular correction has been reduced to $\frac{2}{3}$ of its original value while the near point vergence is now twice the original ocular correction. Estimate the age of the eye.

The only difference is that instead of the increase in age, the estimated age must be found.

On the basis that a 13 year old person can accommodate 13 D, the original amplitude of accommodation enables one to calculate the original age, namely a decrease in amplitude of $+13 - (+9) = +4$ D takes place in $+4.00/+0.25 = 16$ years. Hence, the original age was $13 + 16 = 29$ years. The new age will, therefore, be about $29 + 30 = 59$ years.

Alternatively, one may reason that $+13 - (+1.33) = +11.67$ D are lost in $+11.67/+0.25 = \sim 46$ years. Hence, the new age will be $13 + 46 = 59$ years (approx.).

EXAMPLE 42

At a given moment the ocular correction is equal to the near point vergence, but carries the opposite sign. Twelve years later the amplitude of accommodation is equal to the ocular correction. Calculate the original near point distance.

*Solution 42*

Original situation:

$$B = -K$$

New situation:

$$\text{new}\,Amp = K \qquad \text{(a)}$$

Since the amplitude of accommodation is given by equation (12.5) as

$$Amp = K - B$$

by substitution, the original amplitude was

$$\begin{aligned} Amp &= K - (-K) \\ &= 2K \end{aligned} \qquad \text{(b)}$$

In the course of 12 years the accommodation is reduced by $12 \times (+0.25) = +3.00$ D. From equations (a) and (b) it follows that

$$\text{new } Amp = \frac{1}{2} \text{original } Amp$$

Hence the new amplitude of accommodation is +3.00 D and the original amplitude was +6.00 D. From equation (b),

$$2K = +6.00 \text{ D}$$

or

$$K = +3.00 \text{ D}$$

Substituting this into equation (a),

$$B = -(+3.00) = -3.00 \text{ D}$$

and so the original near point distance was $b = -33.33$ cm.

The variations on this theme are many.

## 13.7 Increasing the vertex distance

As was shown in section 9.4 when the vertex distance is increased the effective power of the spectacle correction at the eye is increased too, that is, the vergence incident at the eye becomes more positive. This is of practical importance in near vision on two accounts.

Firstly, the corrected pre-presbyope will find that by moving his distance correction down his nose he can carry out a near vision task for longer periods without discomfort. Secondly, while wearing his near vision correction the presbyope will be able to see clearly at a distance shorter than the usual working distance when he moves his spectacle frame down his nose.

Equation (9.9) shows, that, in general, for myopic and hyperopic corrections the difference in effectivity of the spectacle lens at the eye is a function of the power of the spectacle lens and the increase in the vertex distance.

Jalie (1977) discusses this and related subjects in greater detail.

# Chapter Fourteen
# Aphakia

## 14.1 Definition

Aphakia (a = not; phakos = lens) is the condition where the crystalline lens is absent from the eye. This is another term coined by Donders (1864).

## 14.2 Introduction

The most common cause of aphakia is an artificial one: the crystalline lens has been removed in view of cataract. The latter condition is present when part of, or the whole lens, has become subject to opacification. When a cataract has formed, the eye's ability to see is interfered with; it may be compared with looking through frosted glass. Depending on the density of the cataract, vision may vary from 'normal' to 'light perception'. In view of this interference with vision, the cataractous lens is often removed by means of an operation. Hence, the eye has become aphakic.

Aphakia can be congenital, or the crystalline lens is indeed present but may have become dislocated as a result of an injury and is now no longer situated perpendicular to the visual axis. Occasionally, the lens is partially dislocated resulting in a simultaneously phakic and aphakic image formation in the eye.

The reader will appreciate that, when an eye is aphakic, it cannot accommodate.

## 14.3 Visual optics

When an eye is aphakic it can no longer be represented by the reduced eye as used this far. To recapitulate, the reduced surface represents the

combination of cornea and crystalline lens. This system is no longer intact in an aphakic eye. The refractive power of the eye is now due solely to the cornea. The position of the cornea with respect to the reduced surface has been discussed in section 4.2. The plane A which passes through the apex of the cornea was shown to be situated $1\frac{2}{3}$ mm in front of the reduced surface (*see Figure 4.3*). In aphakia the axial length of a reduced eye that was originally of standard length, now measures

$$\begin{aligned} k' &= \mathrm{AP} + \mathrm{PM}' \\ &= \mathrm{AP} + k'_o \\ &= +1.67 + 22.22 \\ &= +23.89 \text{ mm} \end{aligned}$$

NOTE

In aphakia distances measured originally with respect to the reduced surface ($k, k', r, f_e, f'_e, d$), will have to be measured from the corneal apex A. Appropriate alteration, by $1\frac{2}{3}$ mm, will be required of the axial length and the vertex distance. This new axial length requires a dioptric length different from that of the standard reduced eye. It may be calculated from equation (5.1):

$$\begin{aligned} K' &= \frac{1333}{+23.89} \\ &= +55.80 \text{ D} \end{aligned}$$

The dioptric power of the cornea of an aphakic eye may be calculated from its radius of curvature. In this example the equivalent corneal surface radius of Gullstrand's Simplified Schematic Eye (No.2), shown in *Table 2.1*, will be used. The power of the eye, which is equal to the power of the cornea, is given by (equation (1.4))

$$\begin{aligned} F_e &= \frac{1333 - 1000}{+7.8} \\ &= +42.69 \text{ D} \end{aligned}$$

The ametropia formula (equation (5.2)) shows that this eye has an ocular correction of

$$\begin{aligned} K &= +55.80 - (+42.69) \\ &= +13.11 \text{ D} \end{aligned}$$

Assuming the vertex distance (*see* section 9.2.2) of the aphakic eye to be 12 mm (hence, the vertex distance of this eye when still phakic, was $13\frac{2}{3}$ mm), the spectacle correction may be calculated as set out below:

$$K = +13.11\text{ D} \longrightarrow k = +7.63\text{ cm}$$
$$d = +1.2 \quad +$$
$$F_{sp} = +11.33 \longleftarrow f'_{sp} = +8.83\text{ cm}$$

The reader will appreciate the relationship between the aphakic eye's refractive power and its first and second focal lengths.

The spectacle magnification is large compared with values for phakic eyes. For the above given example, spectacle magnification is (equation (10.6))

$$SM = \frac{+13.11}{+11.33}$$
$$= +1.157\,1$$

or almost 16%! (*see Figure 15.4(k)*).

The relative spectacle magnification may be calculated from either equation (11.4) or (11.7). The first is very appropriate in this case since it is based on the spectacle magnification and the ratio of the axial lengths, and gives

$$RSM = \left(\frac{+23.89}{+22.22}\right)(+1.157\,1)$$
$$= (+1.075\,2)(+1.157\,1)$$
$$= +1.244\,1$$

The second formula is based on the dioptric powers involved, and gives

$$RSM = \frac{+60.00}{+11.33 + 42.69 - 0.012(+11.33)(+42.69)}$$
$$= \frac{+60.00}{+54.02 - 5.81}$$
$$= \frac{+60.00}{+48.21}$$
$$= +1.244\,6$$

They show a value approaching 25%! It indicates that the retinal image of this corrected aphakic eye is about 25% larger than that of the standard reduced emmetropic eye.

## 14.4 Shape factor

So far all spectacle lenses have been treated as if they were infinitely thin. However, it will be clear that a lens of some +11 D having a diameter sufficiently large to be glazed in a spectacle frame, can no longer be considered as 'thin'. This being so, a further factor needs to be taken into consideration, namely the so-called 'shape factor'. The shape factor has been discussed by several authors, notably by Bennett (1956, 1968a, 1968b) and Jalie (1977). It may be described as the inherent magnification of a lens, and may be summarized as follows.

In section 1.2 the equivalent lens and principal planes of thick lens systems were discussed. Since aphakic spectacle lens corrections must be considered to be thick lenses, the reader might like to refer to this section. One of the points made in that section was that the second principal plane P′ is displaced towards the more positive power component. In fact, both principal planes are displaced in that direction in spectacle lenses (*Figure 14.1*).

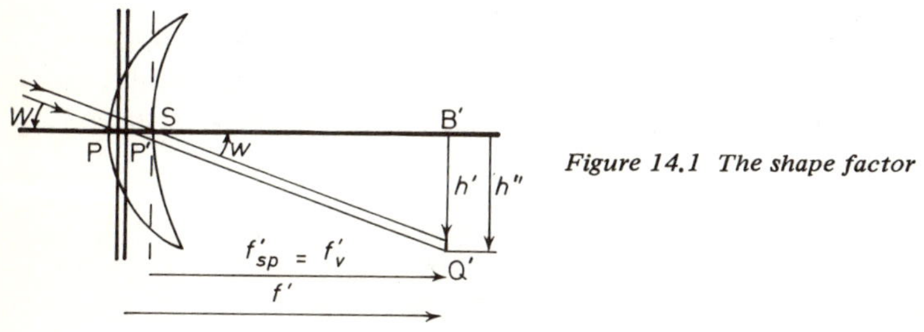


*Figure 14.1 The shape factor*

When a thick lens is treated as 'thin', the image size $h'$ of a distant object is given by the equation

$$h' = -f'_{sp} \tan w$$

with respect to the spectacle plane S. However, the principal planes of a thick, curved spectacle lens lie well in front of the spectacle plane S. Therefore, the second equivalent focal length $f'$ ought to be used in this equation instead of the back vertex focal length $f'_v$ of the spectacle lens (which has been denoted by $f'_{sp}$ up to this point). Consequently, the image size $h''$ due to the thick lens, is given by

$$h'' = -f' \tan w$$

As can be seen from *Figure 14.1*, $h'' > h'$. The difference in size is due to the curving of the lens. The shape factor $S$ may be defined as

$$S = \frac{h''}{h'}$$

$$= \frac{-f' \tan w}{-f'_{sp} \tan w}$$

Since $f'_{sp} = f'_v$,

$$S = \frac{f' \tan w}{f'_v \tan w}$$

$$= \frac{F'_v}{F} \qquad (14.1)$$

According to equation (1.23), the back vertex power $F'_v$ is given by

$$F'_v = \frac{F_1}{1-(t/n)F_1} + F_2$$

$$= \frac{F_1 + F_2 - (t/n)F_1F_2}{1-(t/n)F_1}$$

where $d$ has been substituted by $t$ (denoting the thickness of the lens). The equivalent focal power is given by (equation (1.19))

$$F = F_1 + F_2 - (t/n)F_1F_2$$

where the same substitution has been made.
Hence,

$$S = \frac{F'_v}{F}$$

$$= \frac{[F_1 + F_2 - (t/n)F_1F_2]/[1-(t/n)F_1]}{F_1 + F_2 - (t/n)F_1F_2}$$

$$= \frac{1}{1-(t/n)F_1} \qquad (14.2)$$

In other words: the shape factor depends on the reduced thickness of the lens and the front surface power.

Bennett (1968a) investigated, amongst others, this factor and calculated the shape factor values of typical spectacle lenses. The shape factor of negative spectacle lenses up to −20.00 D varies from 1.01 to

1.00, and may be ignored. However, the shape factor of positive lenses cannot be overlooked. Bennett gave the following values in *Table 14.1.*

TABLE 14.1

| $F'_v = F_{sp}$ (D) | *Back surface power* (D) | *Centre thickness* (mm) | *S* |
|---|---|---|---|
| + 2.00 | − 5.50 | 2.5 | 1.01 |
| 4.00 | 5.00 | 3.6 | 1.02 |
| 6.00 | 4.50 | 4.8 | 1.03 |
| 8.00 | 3.00 | 5.0 | 1.04 |
| 10.00 | 3.00 | 6.0 | 1.05 |
| 12.00 | 3.00 | 6.8 | 1.06 |
| 14.00 | 3.00 | 7.6 | 1.08 |
| 16.00 | 3.00 | 8.4 | 1.10 |
| 18.00 | plano | 8.8 | 1.10 |
| 20.00 | plano | 9.8 | 1.13 |

In view of the high value of the shape factor of around 1.05 for an aphakic spectacle correction, neither the spectacle magnification nor the relative spectacle magnification can be considered on its own, but must be considered in conjunction with the shape factor. Therefore, in aphakia and in other cases requiring high powered positive spectacle lenses, the total spectacle magnification is given by

$$SM \times S$$

and the total relative spectacle magnification by

$$RSM \times S$$

Substituting the values calculated in the previous section, one finds

$$\text{total } SM = (+1.157\,1)(+1.05) = +1.215\,0$$

and

$$\text{total } RSM = (+1.244\,1)(+1.05) = +1.306\,3$$

## 14.5 Conclusion

The large increase in retinal image size in this type of ametropia brings to mind the points made in the chapters on spectacle magnification (chapter 10) and relative spectacle magnification (chapter 11). Its significance in connection with ocular rotations in binocular vision will be discussed in chapter 18.

The above given example is based on an originally emmetropic (phakic) eye. However, if an aphakic eye was originally hyperopic, its ametropia will become even higher when aphakic. An originally myopic eye having been rendered aphakic, will require a less powerful positive correction than an originally emmetropic eye. If a myopic eye were about 13 D ametropic, the removal of the crystalline lens would make it near enough emmetropic. However, apart from the fact that it would no longer be able to accommodate, surgeons are not keen to perform such an operation in view of the possible risks associated with any operation (Rubin, 1967).

A further point should be made, namely, notwithstanding the large retinal image size differences that are present in pairs of eyes where one eye is aphakic, the patient may *not* complain of any problems with binocular single vision. It would appear that there simply is no binocular vision present! This can only be established by means of a clinical trial (Chaston, 1975).

It has often been observed that an aphakic is able to read without a near vision addition. The explanation must be sought in the fact that a near object produces a larger basic retinal image, that the distance correction produces a large spectacle magnification, and that by pushing the spectacle down the nose the effectivity of the lens at the eye increases in positive power. As for the latter point, if a +11 D spectacle correction is moved 1 cm further from the eye, it increases the effectivity at the original spectacle point by +1.2 D (equation (9.12)) to +12.2 D.

# Chapter Fifteen
# Astigmatism

## 15.1 Introduction

Astigmatism in general has been discussed in section 1.3, its classification with respect to the eye in section 3.3, and some clinical aspects in section 3.4.4.

Ocular astigmatism is usually due to a departure from a spherical shape of one or more of the refracting surfaces of the eye: the anterior and posterior corneal surfaces and the anterior and posterior surfaces of the crystalline lens.

It is usually not important to know the cause of ocular astigmatism since the correction by means of astigmatic spectacle lenses follows a standard pattern, at least in most cases. However, it can be of importance in case of a contact lens correction.

## 15.2 The reduced eye

Since the reduced surface is the only refracting surface in the reduced eye, astigmatism can only arise there. This must mean that in an astigmatic reduced eye the radius of curvature of the reduced surface will vary from meridian to meridian, and that there will be a shortest and a longest radius of curvature. As a consequence, there must be two nodal points: the first associated with the shortest and the second with the longest radius. Furthermore, the shortest radius of curvature will give rise to the most powerful meridional power and the longest radius to the least powerful meridional power. These meridional powers are situated *along* the principal meridians of the astigmatism. On the cornea every meridional power will be positive.

Since there is only one axial length each principal meridional power will be associated with an ocular correction of its own. In turn, there will be two far point distances and two far points.

Since accommodation is assumed to be spherical, there will also be two near points, two near point distances, two near point vergences as well as two ranges of accommodation, in every astigmatic eye.

## 15.3 Image formation

### 15.3.1 POINT OBJECT

In astigmatism a point object is not imaged as a point image, but as two focal lines. *Figure 1.25* shows a perspective representation of an astigmatic image formation. Since the image formation is quite complicated, the reader might prefer to consider each principal meridian in turn (*Figure 15.1*). The most powerful meridian is associated with the

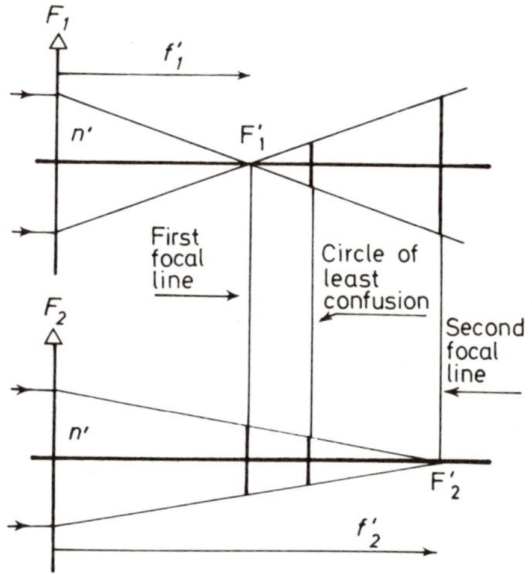


*Figure 15.1 The astigmatic image formation considered along the two principal meridians*

shortest focal length, and the least powerful meridian with the longest focal length. The size of each focal line can be determined from the appropriate couple of similar triangles based on the diameter of the lens, and the image distance. The calculation is basically the same as that of a blur circle (chapter 8).

Situated in the 'interval of Sturm', that is, between the two focal lines, is the 'circle of least confusion' (*Figure 1.25*). This is the point along the axis of the astigmatic system where the astigmatic pencil has a circular section; hence, the use of the word 'circle'. A section taken anywhere

else through an astigmatic pencil at right angles to its axis has the shape of an ellipse (that is, if a focal line is considered to be a special case of an ellipse).

The position of the circle of least confusion (of a distant point object) with respect to the lens is, in *terms of dioptres*, halfway between the two focal lines. (For a proof, *see* Jalie, 1977). Hence, the circle of least confusion's vergence is

$$\frac{F_1 + F_2}{2} \tag{15.1}$$

and the distance of the circle of least confusion from the lens is

$$\frac{2n'}{F_1 + F_2} \tag{15.2}$$

where $F_1$ and $F_2$ represent the powers of the principal meridians.

The diameter of the circle of least confusion may be calculated from similar triangles (*Figure 15.1*). The reader will note that there are several ways of arriving at its value.

NOTE

A point object may be imaged by an astigmatic system as either a line, an ellipse or a circle, but not as a point. Which of these images will be formed will depend on the position of the receiving screen in the astigmatic pencil (*Figure 15.2*).

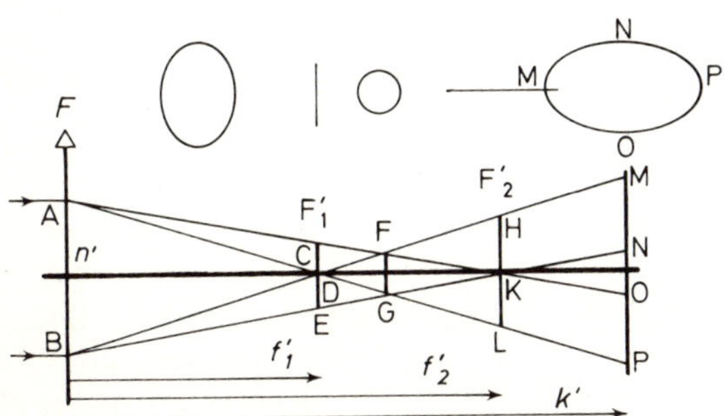


*Figure 15.2 The astigmatic image formation shown in a composite diagram, and sections through the astigmatic pencil (above)*

Readers might like to combine the two parts of *Figure 15.1*, as shown in *Figure 15.2*.

EXAMPLE 43

An astigmatic lens of 5 mm diameter has principal meridional powers of $F_1 = +65.00$ D along 180° and $F_2 = +60.00$ D along 90°SN*. The medium in the image space has a refractive index $n'$ of 4/3. Calculate the size and position of the focal lines and of the circle of least confusion for the case where a parallel pencil of light is incident on the lens (*Figures 15.1* and *15.2*).

*Solution 43*

*Step 1.* Since a pencil of parallel rays of light is incident on the astigmatic lens, the focal lines will be formed in the second focal plane of each principal meridional power. Thus:

| *along 180°* | *along 90°* |
|---|---|
| $f'_1 = \dfrac{1333}{+65.00}$ | $f'_2 = \dfrac{1333}{+60.00}$ |
| $= +20.51$ mm | $= +22.22$ mm |

*Step 2.* The position of the circle of least confusion is (equation 15.2)

$$\frac{2 \times 1333}{+65.00 + 60.00}$$

$$= +21.33 \text{ mm}$$

*Step 3.* The size of the focal lines may be found from the appropriate similar triangles (*Figure 15.2*).

The vertical focal line's length may be calculated from ΔΔCEK and ABK:

$$\frac{\text{CE}}{\text{AB}} = \frac{f'_2 - f'_1}{f'_2}$$

or

$$\text{CE} = \frac{5(22.22 - 20.51)}{22.22}$$

$$= 0.38 \text{ mm}$$

*Standard axis notation (BS 3521:1962).

The horizontal focal line's length may be calculated from ΔΔHLD and ABD:

$$\frac{\text{HL}}{\text{AB}} = \frac{f_2' - f_1'}{f_1'}$$

or

$$\text{HL} = \frac{5(22.22 - 20.51)}{20.51}$$

$$= 0.42 \text{ mm}$$

*Step 4.* The diameter of the circle of least confusion may be calculated from the similar ΔΔFGK and ABK:

$$\frac{\text{FG}}{\text{AB}} = \frac{f_2' - \dfrac{2n'}{F_1 + F_2}}{f_2'}$$

or

$$\text{FG} = \frac{5(22.22 - 21.33)}{22.22}$$

$$= 0.20 \text{ mm}$$

There are alternative triangles that may be used for this purpose. Incorrect values are least likely to be obtained if ΔΔFGK and ABK or ΔΔFGD and ABD are used since this would entail the use of the least number of calculated value (in which errors could have arisen).

EXAMPLE 44

Calculate the image size if a screen (retina) were placed in the image forming pencil of the previous example, at a distance of +23.50 mm from the lens (*Figure 15.2*).

*Solution 44*

The vertical axis of the ellipse formed on the screen may be calculated from ΔΔNOK and ABK:

$$\frac{\text{NO}}{\text{AB}} = \frac{k' - f_2'}{f_2'}$$

or

$$\text{NO} = \frac{5(23.50 - 22.22)}{22.22}$$

$$= 0.29 \text{ mm}$$

The horizontal axis of the ellipse formed on the screen may be calculated from ΔΔMPD and ABD:

$$\frac{\text{MP}}{\text{AB}} = \frac{k' - f_1'}{f_1'}$$

or

$$\begin{aligned}\text{MP} &= \frac{5(23.50 - 20.51)}{20.51} \\ &= 0.73 \text{ mm}\end{aligned}$$

This problem may also be solved by means of equation (8.2), used twice. The dioptric length is $K' = +56.72$ D. Since the object vergence is zero, the image vergences will have the powers of the principal meridians. Hence,

$$\begin{aligned}\text{NO} &= \frac{5(+56.72 - 60.00)}{+56.72} \\ &= 0.29 \text{ mm}\end{aligned}$$

$$\begin{aligned}\text{MP} &= \frac{5(+56.72 - 65.00)}{+56.72} \\ &= 0.73 \text{ mm}\end{aligned}$$

NOTE
The focal line lies always along the axis of the cylinder.

### 15.3.2 EXTENDED OBJECT

An extended object may be considered to consist of a very large number of point objects that are aligned to form a certain shape. Consequently, it is only necessary to decide the dimensions of the image of one of these many point objects, and to build up the extended object's image on that basis.

If, for example, a distant letter were imaged by the optical system of example 44, each single image point would be an oval of 0.73 × 0.29 mm with the longer axis horizontal (*see Figure 15.2*). The images thus formed of a letter T and V are diagrammatically represented in *Figure 15.3(a)*. A photographed image is shown in *Figure 15.4(f)*. If the astigmatic system were rotated through 60°, the focal lines and blur ovals

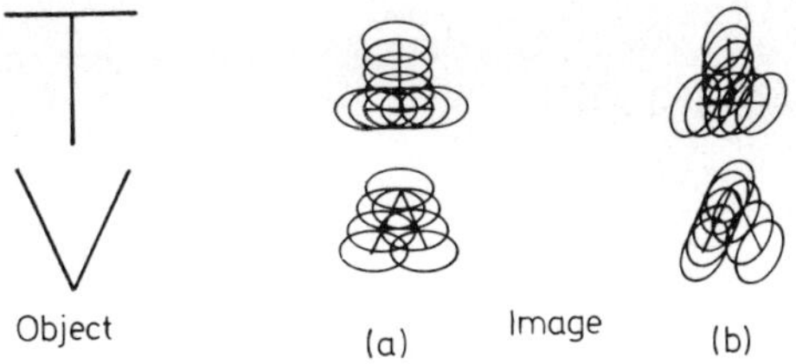

Object

(a) Image (b)

*Figure 15.3 The effect of rotating an astigmatic system through 60°, on the blurred image formation of an extended object*

would also rotate through 60° (*Figure 15.3*(*b*)). Compare also other photographs of *Figure 15.4*.

## 15.4 Retinal image size

By introducing a refractive index $n'$ of 4/3 in the image space in examples 43 and 44, image sizes in reduced astigmatic eyes have been calculated.

Here follows another example:

EXAMPLE 45

A reduced eye has principal meridional powers of $F_{e1}$ = +62.50 D along 120° and $F_{e2}$ = +58.00 D along 30°. The eye's axial length is

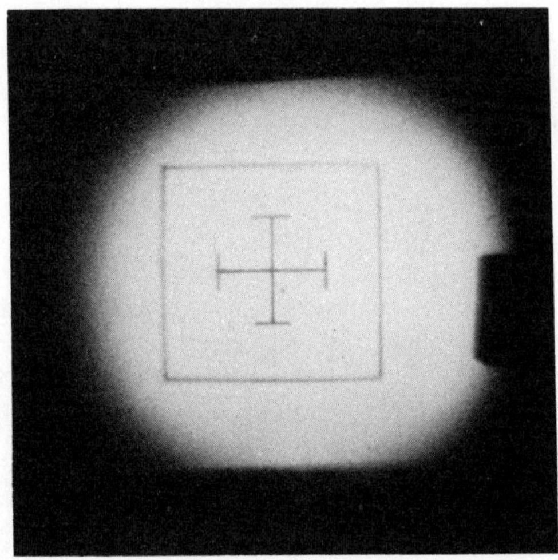

*Figure 15.4 The following photographs are 'retinal' images formed in the model eye built and described by Moss (1976, 1977)*

*Figure 15.4(a) The retinal image formed in the emmetropic model eye*

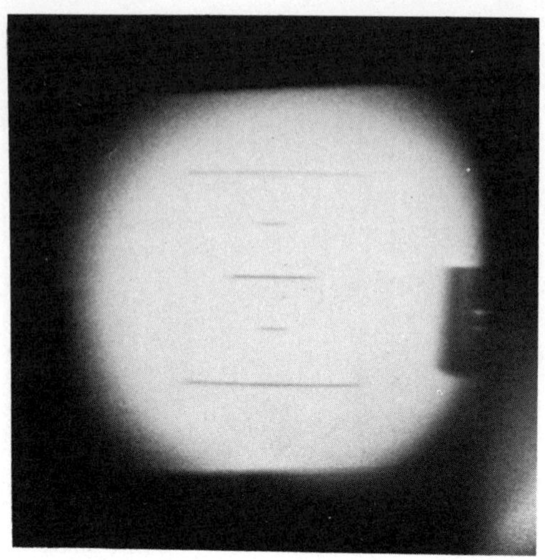

*Figure 15.4(b) The retinal image formed in the uncorrected astigmatic model eye. Spectacle correction – 7.50 DC axis 90°*

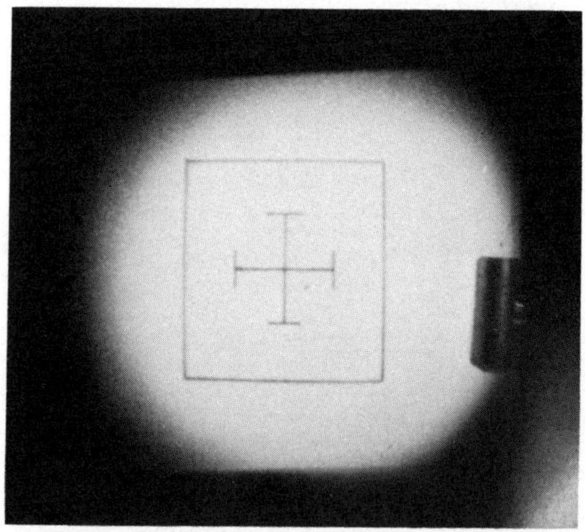

*Figure 15.4(c) The retinal image formed in the model eye of Figure 15.4(b), when corrected*

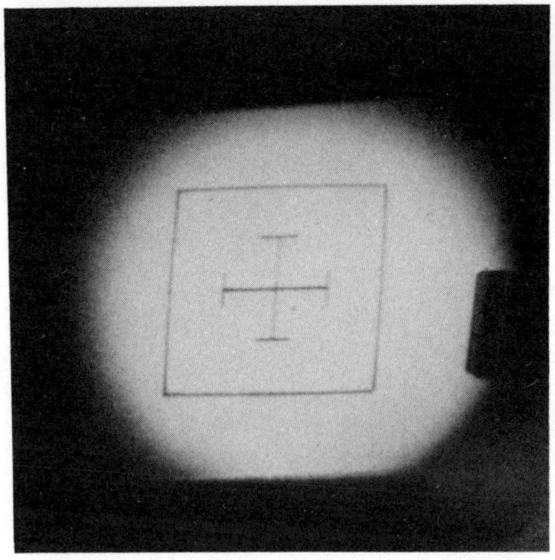

*Figure 15.4(d) The retinal image formed in the corrected astigmatic model eye, of spectacle correction —7.50 DC axis 60°*

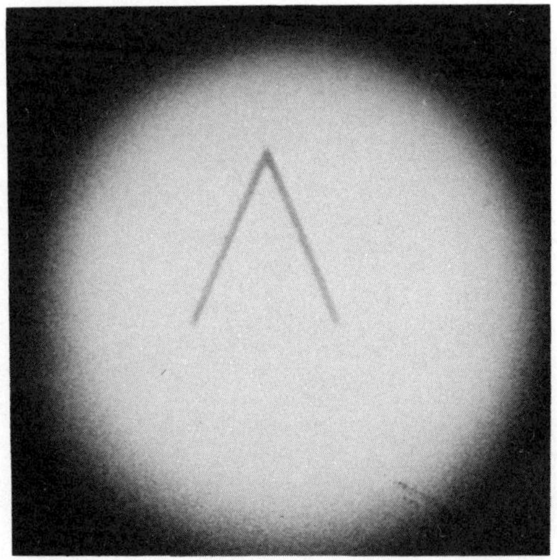

*Figure 15.4 (e) The retinal image in the emmetropic model eye, of a V-shaped object*

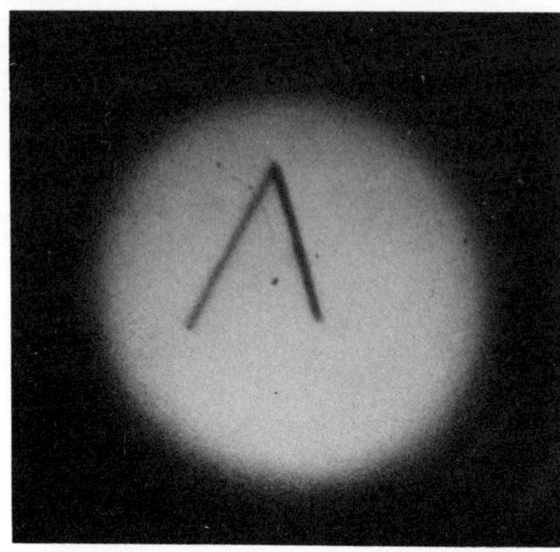

*Figure 15.4 (f) The retinal image of the same object, in the uncorrected astigmatic model eye. Spectacle correction −7.50 DC axis 60°. Compare with* Figure 15.4(b)

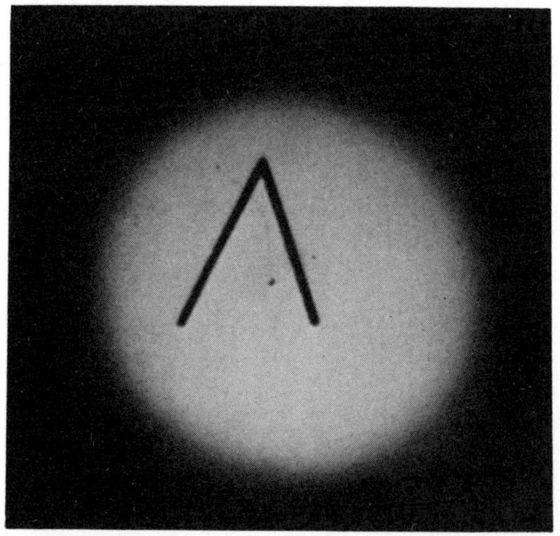

*Figure 15.4 (g) The retinal image of the V-shaped object, in the corrected astigmatic model eye. Spectacle correction −7.50 DC axis 60°*

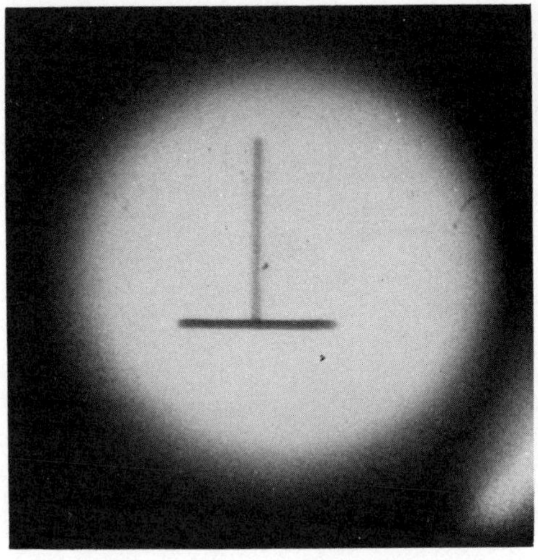

*Figure 15.4 (h) The retinal image of a T-shaped object in the uncorrected astigmatic model eye. Spectacle correction −7.50 DC axis 90°*

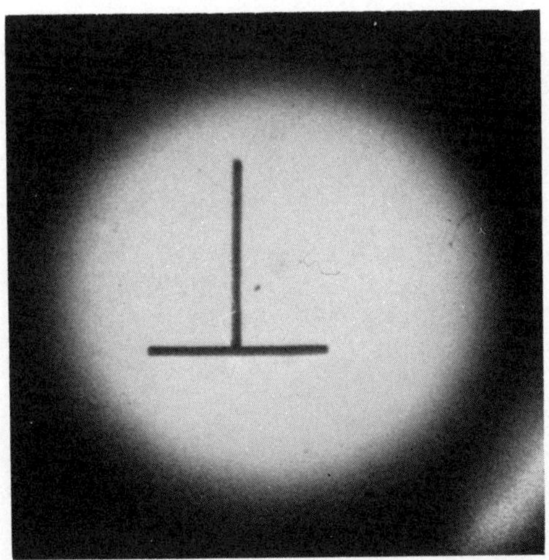

*Figure 15.4 (j) The retinal image of the T-shaped object in the same model eye when corrected. For comparison, measure the length of each limb in this and in the previous photograph*

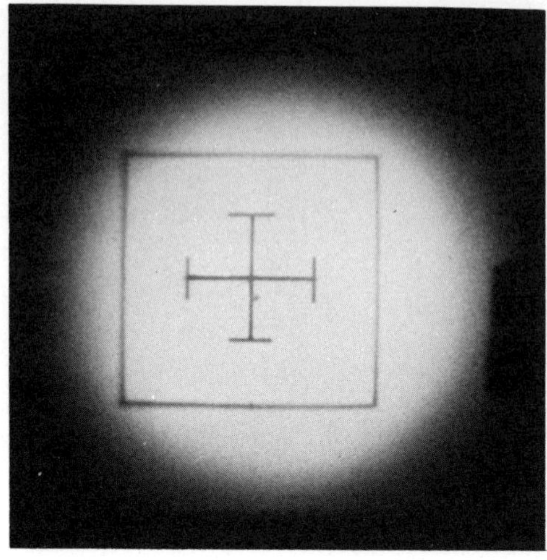

*Figure 15.4 (k) The retinal image formed in a corrected aphakic model eye. Spectacle correction +8.75 DS at 10 mm vertex distance. For comparison, measure the length of a limb in this photograph and in* Figure 15.4(a)

+22.50 mm. It looks at an object which is 2 cm high and placed 2 m in front of the eye. Calculate the dimensions of the retinal image for a pupil diameter of 6 mm (*Figure 15.5*).

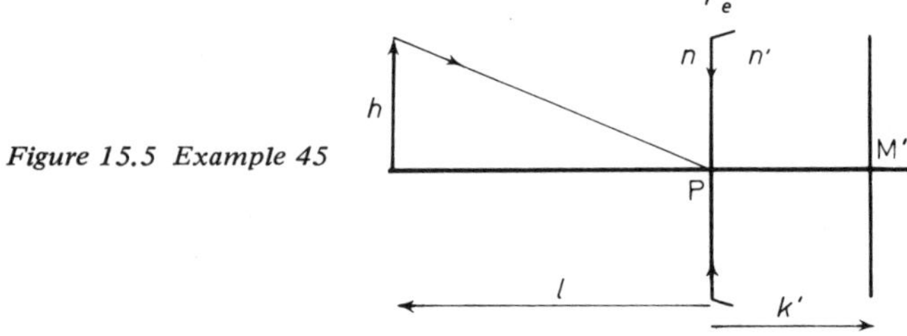


*Figure 15.5 Example 45*

*Solution 45*

*Step 1.* The object vergence is given by

$$L = \frac{1}{-2}$$

$$= -0.50 \text{ D}$$

*Step 2.* The basic retinal image size $h'$ may be calculated as follows:

$$\frac{-h'}{k'} = -\frac{3}{4} \times \frac{h}{l}$$

or

$$h' = \frac{3}{4} \times \frac{k'h}{l}$$

$$= -\frac{3 \times 22.50 \times 20}{4 \times 2000}$$

$$= -\frac{1350}{8000}$$

$$= -0.17 \text{ mm}$$

Add this information to the diagram (*Figure 15.6*).

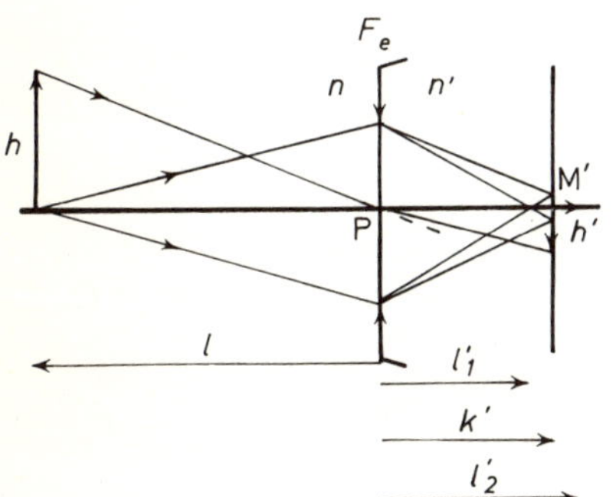


*Figure 15.6 Example 45; a method of representing the basic retinal image size and blur oval dimensions in one diagram*

*Step 3.* The image vergence may be calculated from the conjugate foci formula:

| *along 120°* | *along 30°* |
|---|---|
| $L'_1 = -0.50 + 62.50$ | $L'_2 = -0.50 + 58.00$ |
| $= +62.00$ D | $= +57.50$ D |

*Step 4.* The image distance is given by (equation (1.11))

| | |
|---|---|
| $l'_1 = \dfrac{1333}{+62.00}$ | $l'_2 = \dfrac{1333}{+57.50}$ |
| $= +21.50$ mm | $= +23.18$ mm |

Add this information to the diagram (*Figure 15.6*).

*Step 5.* Calculate the blur oval dimensions on the retina. The lengths of the axes are (equation (8.1)):

$$j = \frac{6(22.50 - 21.50)}{21.50} \qquad j = \frac{6(23.18 - 22.50)}{23.18}$$

$$= 0.28 \text{ mm measured along } 30^\circ \text{ SN} \qquad = 0.18 \text{ mm measured along } 120^\circ \text{ SN}$$

*Step 6.* Draw a diagram to scale, of the retinal image (*Figure 15.7*).

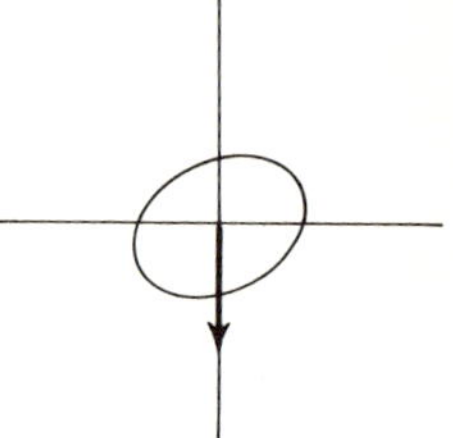

*Figure 15.7 Example 45; a simplified representation of the retinal image showing the basic retinal image and the blur oval to scale*

NOTE

The image is inverted, and every image point will be imaged as the oval shown centred on the foot of the image (arrow).

The blur oval dimensions on the retina may also be calculated from equation (8.2) using vergences rather than distances.

## 15.5 Accommodation and the astigmatic eye

Accommodation is generally accepted to be a phenomenon whereby the power of an eye increases equally in all meridians. In other words: the spherical power of the eye is increased when accommodation takes place.

In visual optics calculations the spherical nature of accommodation is usually taken for granted. Astigmatic accommodation, however, has been the subject of several investigations. Notwithstanding these (Le Grand, 1964; Obstfeld, 1964; Duke-Elder, 1970; Bannon 1972; to mention but a few), it has remained a matter of some controversy.

If the spherical nature of accommodation is accepted, calculations will not present problems that have not been dealt with in earlier chapters. For the purpose of an example, consider the following exercise:

EXAMPLE 46

The far points of an astigmatic eye are found to be situated at +50 cm

and at −200 cm. The amplitude of accommodation is 5.00 D. Find the near point distances.

*Solution 46*

*Step 1.* Calculate the ocular correction of each principal meridian:

*1st meridian*

$$K = \frac{100}{+50.00} = +2.00 \text{ D}$$

*2nd meridian*

$$K = \frac{100}{-200.00} = -0.50 \text{ D}$$

*Step 2.* The near point vergences may be calculated from equation (12.5):

$$B = +2.00 - (+5.00) = -3.00 \text{ D}$$

$$B = -0.50 - (+5.00) = -5.50 \text{ D}$$

*Step 3.* The near point distances are (equation (12.4)):

$$b = \frac{100}{-3.00} = -33.33 \text{ cm}$$

$$b = \frac{100}{-5.50} = -18.18 \text{ cm}$$

However, another problem could be posed: how much accommodation is an uncorrected astigmatic eye going to exert when an object is situated between its far point and its near point? Is it going to exert sufficient accommodation to bring its second focal line on the retina, or the circle of least confusion, or perhaps the first focal line? There would not appear to be a clear cut answer to this question. At least one author (Fry, 1940) implies that any one of the above mentioned possibilities may be chosen, or that an arbitrary point within the interval of Sturm may be chosen to be imaged on the retina. Further considerations lead into the province of physiological optics which lies outside the scope of this book.

In visual optics one is usually satisfied when the amount of uncorrected astigmatism at the reduced surface of an eye which looks through an astigmatic distance correction at a near object, has been determined. This will be discussed in section 16.4.

# Chapter Sixteen
# Correction of Astigmatism

## 16.1 Introduction

In the previous chapter it was explained that an astigmatic eye has two far points. This should never be forgotten when considering the correction of astigmatism: the main point to remember is that the correction of the two principal meridians must be calculated *separately.*

## 16.2 Spectacle correction

The calculation of the spectacle correction of an astigmatic eye is basically the same as that of a spherically ametropic eye. The only difference is that, since an astigmatic eye has two far points, the calculation must be made with respect to each far point. Consider the following example:

EXAMPLE 47

The principal meridional powers of a refractively ametropic reduced eye are +66.00 D along 55° and +61.00 D along 145°. Calculate the spectacle correction at a vertex distance of 15 mm.

*Solution 47*

*Step 1.* The ocular correction may be calculated from the ametropia formula (equation (5.4)):

| *55° meridian* | *145° meridian* |
|---|---|
| $K = +60.00 - (+66.00)$ | $K = +60.00 - (+61.00)$ |
| $\quad = -6.00$ D | $\quad = -1.00$ D |

*Step 2.* The far point distances are (equation (5.7)):

$$k = \frac{100}{-6.00} = -16.67 \text{ cm} \qquad k = \frac{100}{-1.00} = -100.00 \text{ cm}$$

*Step 3.* Following the step-along method, the spectacle correction in the two meridians is found to be:

$$\begin{array}{rcl} k & = & -16.67 \text{ cm} \\ d & = & +1.5 \quad + \\ \hline f'_{sp} & = & -15.17 \longrightarrow F_{sp} \\ & & = -6.59 \text{ D} \end{array} \qquad \begin{array}{rcl} k & = & -100.00 \text{ cm} \\ d & = & +1.5 \quad + \\ \hline f'_{sp} & = & -98.50 \longrightarrow F_{sp} \\ & & = -1.02 \text{ D} \end{array}$$

*Step 4.* The theoretical spectacle prescription reads

−6.59 DC along 55°/−1.02 DC along 145°

which may be transposed into sphero-cylindrical form as appropriate.

NOTE

The ocular astigmatism is +66.00 − (+61.00) = +5.00 D. The spectacle lens astigmatism (for distance vision) is −6.59 − (−1.02) = −5.57 Note that there is more than ½ D difference between ocular and spectacle astigmatism, and that they carry opposite signs since the one must neutralize the other.

When the vertex distance of a spectacle correction is altered, the amount of spectacle astigmatism alters too:

EXAMPLE 48

A distance correction reads −1.00 DS/−4.00 DC × 90° at 10 mm. The spectacle lens is, in fact, worn at 14 mm. Calculate the new amount of spectacle astigmatism.

*Solution 48*

*Step 1.* Rewrite the prescription in crossed cylinder form:

−1.00 DC × 180/−5.00 DC × 90

*Step 2.* Calculate the new second focal length of each power. Since the vertex distance is increased by 4 mm, these focal lengths alter to:

*axis 180°*

$$\begin{array}{llll} \text{old} & f'_{sp} = -100.00 \text{ cm} & \longleftarrow & F_{sp} = -1.00 \text{ D} \\ & \phantom{f'_{sp} = } \underline{\phantom{-10}0.4\phantom{0 cm}} & + & \\ \text{new} & f'_{sp} = -99.60 & & \end{array}$$

*axis 90°*

$$\begin{array}{llll} \text{old} & f'_{sp} = -20.00 \text{ cm} & \longleftarrow & F_{sp} = -5.00 \text{ D} \\ & \phantom{f'_{sp} = } \underline{\phantom{-1}0.4\phantom{0 cm}} & + & \\ \text{new} & f'_{sp} = -19.60 & & \end{array}$$

*Step 3.* The new powers in the principal meridians are now

$$F_{sp} = -1.00 \text{ D} \qquad\qquad F_{sp} = -5.10 \text{ D}$$

*Step 4.* The spectacle astigmatism is now

$$-1.00 - (-5.10) = +4.10 \text{ D}$$

NOTE

The original amount of spectacle astigmatism was 4.00 D. At the new vertex distance there is only an insignificantly different amount. However, if the original astigmatism had been larger, or if the difference in vertex distance had been larger, or if the least powerful meridian had been stronger, a larger difference in spectacle astigmatism would have been found.

## 16.3 Retinal image size

The retinal image size of a distant object may be calculated in the same way as for a case of spherical ametropia, with this difference that for each principal meridian a separate calculation must be carried out. Consider the following case:

EXAMPLE 49

An astigmatic eye is corrected for distance vision by a lens of +5.00 DS/+4.00 DC × 45° at 12 mm. It looks at a distant square the diagonals of which subtend 1° at the eye. Calculate the retinal image

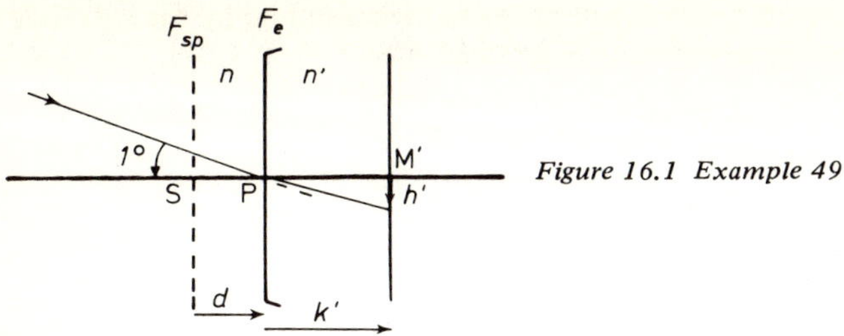


*Figure 16.1 Example 49*

size of the object in the knowledge that the eye's axial length is +23.00 mm (*Figure 16.1*).

*Solution 49*

*Step 1.* Calculate the basic retinal image size of the diagonals (equation (4.3)):

$$h' = -23.00 \times \frac{3}{4} \tan 1°$$

$$= -23.00 \times \frac{3}{4} \times 0.017\,5$$

$$= -0.30 \text{ mm}$$

*Step 2.* Calculate the spectacle magnification for each principal meridian. In order to be able to do this the far point distance has to be calculated first:

| *45° meridian* | *135° meridian* |
|---|---|
| $f'_{sp}$ = +20.00 cm | $f'_{sp}$ = +11.11 cm |
| $d$ = +1.2 − | $d$ = +1.2 − |
| $k$ = +18.80 | $k$ = +9.91 |

The spectacle magnification for each meridian may be calculated from equation (10.5):

$$SM = \frac{+20.00}{+18.80} = +1.063\,8 \qquad SM = \frac{+11.11}{+9.91} = +1.121\,1$$

*Step 3.* To calculate the retinal image dimension along each diagonal

the basic retinal image size has to be multiplied by the spectacle magnification of each principal meridian, thus:

$$h'' = h' \times SM \qquad\qquad h'' = h' \times SM$$
$$= -0.30(+1.063\,8) \qquad\qquad = -0.30(+1.121\,1)$$
$$= -0.32 \text{ mm} \qquad\qquad = -0.34 \text{ mm}$$

Alternatively:
*Step 1.* Calculate the far point image size for each principal meridian (*Figure 16.2*) (equation (10.3)):

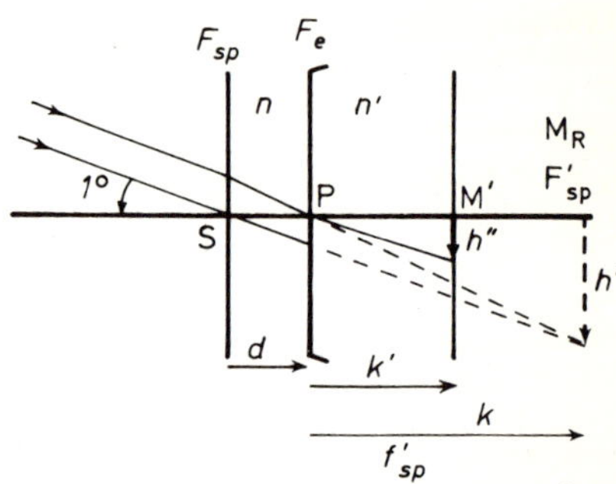


*Figure 16.2 Example 49; alternative approach*

*45° meridian* | *135° meridian*

$$h' = -f'_{sp} \tan 1° \qquad\qquad h' = -f'_{sp} \tan 1°$$
$$= -(+20.00)(0.017\,5) \qquad\qquad = -(+11.11)(0.017\,5)$$
$$= -0.35 \text{ cm} \qquad\qquad = -0.19 \text{ cm}$$

*Step 2.* To find the retinal image size, the far point image size may be multiplied by the magnification of the eye's optical system. To obtain the necessary data, proceed as follows:

The ocular correction may be calculated from the far point distance found in Step 2 above:

$$K = +5.32 \text{ D} \qquad\qquad K = +10.09 \text{ D}$$

The dioptric length of the eye may be calculated from equation (5.1):

$$K' = \frac{1333}{+23.00}$$
$$= +57.96 \text{ D}$$

The magnification of the eye's optical system is given by equation (6.3):

$$m = \frac{+5.32}{+57.96} \qquad\qquad m = \frac{+10.09}{+57.96}$$

$$= +0.091\,8 \qquad\qquad = +0.174\,1$$

The retinal image dimensions are now

$$h'' = -0.35(+0.091\,8) \qquad\qquad h'' = -0.19(+0.174\,1)$$

$$= -0.032 \text{ cm} = -0.32 \text{ mm} \qquad\qquad = -0.033 \text{ cm} = -0.33 \text{ mm}$$

Now consider what has happened to the shape of the retinal image

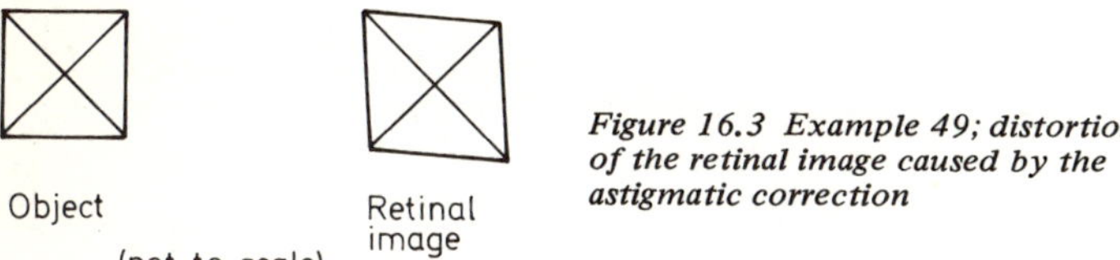


*Figure 16.3 Example 49; distortion of the retinal image caused by the astigmatic correction*

(*Figure 16.3*). The object was a square, the diagonals of which subtended 1° at the eye. The retinal image of the object is no longer a square since the diagonals are no longer of equal length.

NOTE

The reader might like to remember that the larger magnification of the retinal image occurs along the *axis* of the least positive or most negative cylindrical component of the *correcting* lens. This is so because the most positive power produces the higher magnification and acts along the power meridian. The latter coincides with the axis of the least powerful cylindrical component.

## 16.4 Accommodation and presbyopia

In the previous section a distant object was considered. Hence, accommodation could remain relaxed. Now the effect of an astigmatic distance correction used for near vision, will be discussed.

EXAMPLE 50

An object is situated 25 cm in front of a distance correction of

−5.00 DS/−5.00 DC × 30° at 15 mm. Calculate the spectacle and ocular accommodation required.

*Solution 50*

*Step 1.* The spectacle accommodation is calculated according to equation (12.11):

$$A_s = -L_s$$

$$= -\frac{100}{-25.00}$$

$$= +4.00\text{ D}$$

NOTE

This value applies to any meridian.

*Step 2.* The ocular accommodation should be calculated for the two principal meridians. However, the object vergence is common to any meridian. Thus

$$L_s = \frac{100}{-25.00}$$

$$= -4.00\text{ D}$$

Using the step-along method for the two principal meridians, the vergence incident on the eye is

*30° meridian*

$$\begin{array}{lll} L_s = -4.00\text{ D} & & \\ F_{sp} = -5.00 & + & \\ \hline L_s' = -9.00 & \longrightarrow & l_s' = -11.11\text{ cm} \\ & & d = +1.5 \quad - \\ \cline{3-3} L = -7.93 & \longleftarrow & l = -12.61 \end{array}$$

*120° meridian*

$$\begin{array}{lll} L_s = -4.00\text{ D} & & \\ F_{sp} = -10.00 & + & \\ \hline L_s' = -14.00 & \longrightarrow & l_s' = -7.14\text{ cm} \\ & & d = +1.5 \quad - \\ \cline{3-3} L = -11.57 & \longleftarrow & l = -8.64 \end{array}$$

However, the vergence required to be incident on the reduced surface, is:

*30° meridian*

$$F_{sp} = -5.00\text{ D} \longrightarrow f'_{sp} = -20.00\text{ cm}$$
$$d = +1.5 \quad -$$
$$K = -4.65 \longleftarrow k = -21.50$$

*120° meridian*

$$F_{sp} = -10.00\text{ D} \longrightarrow f'_{sp} = -10.00\text{ cm}$$
$$d = +1.5 \quad -$$
$$K = -8.70 \longleftarrow k = -11.50$$

The accommodation required in each of these meridians is (equation (12.3)):

*30° meridian*

$$A = -4.65 - (-7.93)$$
$$= +3.28\text{ D}$$

*120° meridian*

$$A = -8.70 - (-11.57)$$
$$= +2.87\text{ D}$$

The difference in accommodation required is +3.28 – (+2.87) = +0.41 Since accommodation is supposed to be of a spherical nature, one is faced with a problem. In practice, however, this does not usually appear to cause difficulties. This is likely to be due to the fact that any blurring of the retinal image is reduced by a reduction of the pupil diameter which occurs during accommodation, and by the depth of field (section 13.5) which is further increased by the smaller pupil diameter. Hence, in visual optics only the amount of uncorrected ocular astigmatism (in the last example amounting to 0.41 D) in near vision through a distance correction, is considered.

In the very few cases where uncorrected astigmatism during near vision does cause problems, a separate near vision correction, or special bifocal lenses, having a cylinder power and/or axis different from that of the distance correction, may be prescribed. Further details about this type of astigmatism referred to as 'near vision astigmatism' may be found elsewhere (Fletcher, 1952; Obstfeld, 1964; Jalie, 1977).

In the presence of presbyopia, an addition for near vision will be required. The theoretical near vision addition is calculated in the same way as the addition for a spherical distance correction (section 13.4).

If the near object cannot be imaged on the retina because of an insufficient amplitude of accommodation caused by presbyopia, the

blurred retinal image size is calculated in the same way as for any other blurred retinal image.

## 16.5 Aphakia

Astigmatism in an aphakic eye can only be due to corneal astigmatism. The amount of this astigmatism is, therefore, determined by the radii of curvature of the principal meridians of the cornea.

The correction of astigmatism in aphakia follows the same pattern as that of the phakic astigmatic eye.

## 16.6 The corrected astigmatic eye as a telescopic system

The following discussion follows that of section 12.11 where a corrected ametropic eye was treated as if it were a telescopic system. A corrected astigmatic eye may be treated in the same way (Reiner, 1965a, 1965b, 1966, 1972; Obstfeld, 1976b); the system consists of an astigmatic spectacle correction and an imaginary astigmatic contact lens (the ametropia) which represents the telescopic system, and is followed by an emmetropic eye. Telescopes producing different amounts of magnification in mutually perpendicular meridians as a result of the use of an astigmatic objective and an astigmatic eye piece whose axes are parallel, are known as anamorphotic systems. Owing to the different magnifications, distortion is produced. The distortion noticed by a patient who has previously not been corrected with an astigmatic lens, is due to the anamorphotic properties of the combination astigmatic correction–astigmatic eye.

Assuming the two principal meridians of the astigmatism to lie along the horizontal and vertical meridian, the magnification will vary from $M_y$ along the vertical to $M_z$ along the horizontal meridian. Any point of a distant object will then be subject to the two magnifying powers. Consider the point B of a distant circular object (*Figure 16.4*) whose



*Figure 16.4 Distortion caused by an anamorphotic system*

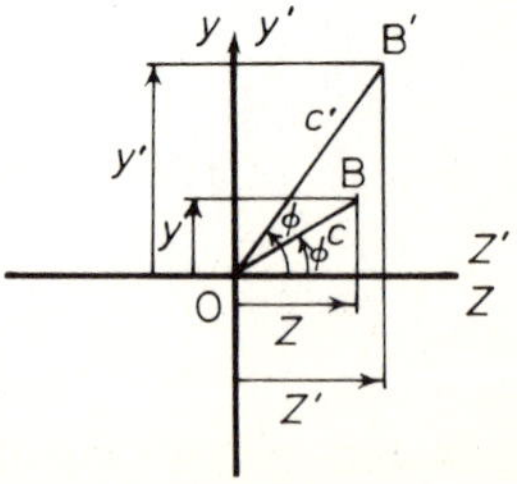

coordinates are $y$ along the vertical and $z$ along the transverse (horizontal) meridian. The coordinates $y'$ and $z'$ of the image B$'$ of point B, are subject to different amounts of magnification. Consequently

$$\left.\begin{aligned} y' &= M_y y \\ \text{and} \quad z' &= M_z z \end{aligned}\right\} \quad (16.1)$$

If $c$ represents the linear distance between B and the optical centre 0 of the lens, and $\phi$ the angle between the transverse axis and B measured at the optical centre, *Figure 16.4* shows that

$$\left.\begin{aligned} y &= c \sin \phi \\ \text{and} \quad z &= c \cos \phi \end{aligned}\right\} \quad (16.2)$$

where the optical axis runs through the point 0 and perpendicular to the plane of the paper.

Similarly,

$$\left.\begin{aligned} y' &= c' \sin \phi' \\ \text{and} \quad z' &= c' \cos \phi' \end{aligned}\right\} \quad (16.3)$$

where $c'$ is the linear distance between the image B$'$ and the optical centre 0, and $\phi'$ is the angle between the transverse axis and the point B$'$ measured at the optical centre.

By substituting equations (16.2) and (16.3) into equation (16.1), one finds

$$c' \sin \phi' = M_y \, c \sin \phi \quad (16.4)$$

and

$$c' \cos \phi' = M_z \, c \cos \phi \quad (16.5)$$

Dividing equation (16.4) by equation (16.5), one obtains

$$\tan \phi' = \frac{M_y}{M_z} \tan \phi \quad (16.6)$$

The factor $M_y/M_z$ of this expression represents the anamorphotic factor.

A property of the anamorphotic image formation is that lines in the object plane that make an angle $\phi$ with the horizontal plane, are imaged at an angle $\phi'$.

The relationship of the length $c'$ measured from the image point B′, and the length $c$ measured from the object point B, to the optical axis, is derived as follows (*Figure 16.4*):

$$c' = \sqrt{(y'^2 + z'^2)}$$

which may be written as

$$c' = \sqrt{(M_y^2\, y^2 + M_z^2\, z^2)}$$

in view of equation (16.1). By substituting equation (16.2),

$$c' = \sqrt{(M_y^2\, c^2 \sin^2\phi + M_z^2\, c^2 \cos^2\phi)}$$

$$= c\sqrt{(M_y^2 \sin^2\phi + M_z^2 \cos^2\phi)} \qquad (16.7)$$

When the point B lies on the vertical meridian, that is when $\phi = 90°$, equation (16.7) reduces to

$$c' = c\,M_y \qquad (16.8)$$

since $\sin 90° = 1$ and $\cos 90° = 0$. Similarly, when point B lies on the horizontal meridian, equation (16.7) reduces to

$$c' = c\,M_z \qquad (16.9)$$

since $\sin 0° = 0$ and $\cos 0° = 1$.

This means that a circular object will be imaged as an ellipse having a vertical axis given by equation (16.8) and a horizontal axis given by equation (16.9). A square shaped object, of side $y$, will be deformed and imaged as a rectangle having a vertical side length according to equation (16.8) and a horizontal side length according to equation (16.9), or the ratio of the sides will be equal to the anamorphotic factor.

The lengths $c$ and $c'$ contain an angle $\theta$ measured at the point 0, where

$$\theta = \phi' - \phi$$

Now

$$\tan\theta = \tan(\phi' - \phi)$$

$$= \frac{\tan\phi' - \tan\phi}{1 + \tan\phi' \tan\phi}$$

In view of equation (16.6) this may be written as

$$\tan\theta = \frac{\dfrac{M_y}{M_z}\tan\phi - \tan\phi}{1 + \dfrac{M_y}{M_z}\tan\phi\tan\phi}$$

$$= \frac{\tan\phi\left(\dfrac{M_y}{M_z} - 1\right)}{1 + \dfrac{M_y}{M_z}\tan^2\phi} \tag{16.10}$$

Thus angle $\theta = 0°$ when the angle $\phi$ is either 0° or 90°, since $\tan 0° = 0$ and $\tan 90° = \infty$. This means that the vertical and horizontal lines of an object will be imaged as vertical and horizontal lines. However, when the principal meridians of an anamorphotic system lie obliquely, vertical and horizontal lines of the object will be imaged as oblique lines, thus giving rise to distortion.

# Chapter Seventeen
# Newton's Equation

## 17.1 Introduction

This equation (equation (1.14)) is not usually applied to solve visual optics problems. As has been shown (Obstfeld, 1972), it can well be used for this purpose, but need not offer an advantage over the classical type of solution (Bennett, 1972).

## 17.2 Visual optics

Newton's equation, $xx' = ff'$, may be written for the purposes of visual optics as (*Figure 17.1*)

$$xx' = f_e f_e' \qquad (17.1)$$

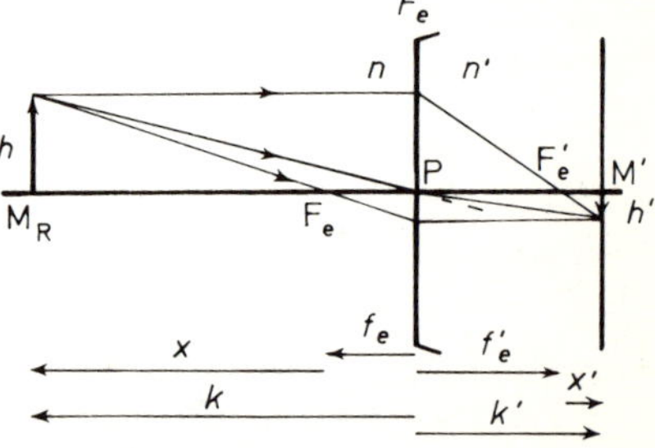


*Figure 17.1 Newton's equation applied to an ametropic eye*

where $x$ = the distance between the eye's first principal focus and the object,
$x'$ = the distance between the eye's second principal focus and the image.

In general, Newton's equation cannot be used when object or image is situated at infinity, for either distance $x$ or $x'$ would become infinitely

large. In short, the equation cannot be used in solving simple problems involving emmetropic eyes.

In the manipulation of optical formulae distances are often replaced by reduced vergences. Newton's equation may thus be altered as follows:

$$\frac{n}{X} \times \frac{n'}{X'} = \frac{-n}{F_e} \times \frac{n'}{F_e}$$

or

$$\frac{nn'}{XX'} = -\frac{nn'}{F_e^2}$$

or

$$nn'F_e^2 = -nn'XX'$$

After dividing both sides of the equation by $(nn')$ and rearranging it, one finds

$$XX' = -(F_e)^2 \qquad (17.2)$$

For the reduced eye of standard power one may write

$$\begin{aligned} XX' &= -(F_o)^2 \\ &= -3\,600.00 \text{ D} \qquad (17.3) \end{aligned}$$

From *Figure 17.1* it can be seen that

$$x = k - f_e \qquad (17.4)$$

and

$$x' = k' - f'_e \qquad (17.5)$$

When the far point distance $k$ is long relative to the first focal length of the eye, $x$ approximates to $k$ and so equation (17.2) becomes

$$KX' = -(F_e)^2 \qquad (17.6)$$

EXAMPLE 51

Calculation the axial length of an axially myopic eye of 1.50 D ocular correction.

*Solution 51*
Following the accurate method:
*Step 1.* Equation (17.4) may be written as

$$x = k - f_o$$

and so

$$x = \frac{1000}{-1.50} - (-16.67)$$

$$= -666.67 + 16.67$$

$$= -650.00 \text{ mm}$$

*Step 2.* Since

$$X = \frac{n}{x}$$

the vergence $X$ is

$$X = \frac{1000}{-650.00}$$

$$= -1.54 \text{ D}$$

*Step 3.* From equation (17.3), the reduced vergence $X'$ is

$$X' = \frac{-3\,600.00}{-1.54}$$

$$= +2\,337.66 \text{ D}$$

Since

$$x' = \frac{n'}{X'}$$

the distance $x'$ is

$$x' = \frac{1333}{+2\,337.66}$$

$$= +0.57 \text{ mm}$$

*Step 4.* From equation (17.5) the axial length is given in this case by

$$k' = x' + f'_o$$

and hence,

$$k' = +0.57 + 22.22$$
$$= +22.79 \text{ mm}$$

Following the approximate method:
*Step 1.* Equation (17.6) may here be written as

$$KX' = -(F_o)^2$$

or

$$X' = \frac{-(F_o)^2}{K}$$

$$= \frac{-3\,600.00}{-1.50}$$

$$= +2\,400.00 \text{ D}$$

Therefore, the distance $x'$ is given by

$$x' = \frac{1333}{+2\,400.00}$$

$$= +0.56 \text{ mm}$$

*Step 2.* Equation (17.5) may be written in this case as

$$k' = x' + f'_o$$

and gives

$$k' = +0.56 + 22.22$$
$$= +22.78 \text{ mm}$$

The limit of application of the approximate formula (equation (17.6)) may be placed at 3.00 D of axial ametropia where the formula gives an error for $x'$ of 0.05 mm which is equivalent to about 0.12 D.

It will be clear (*Figure 17.1*) that such an approximation cannot apply inside the eye where all distances are small.

## 17.3 Accommodation

If accommodation is exerted, equation (17.2) becomes

$$XX' = -(F_e + A)^2$$
$$= -(F_a)^2 \quad (17.7)$$

If the near object distance is still relatively long with respect to the shortened (as a result of accommodation) first focal length, equation (17.6) may be altered to give the approximate formula

$$BX' = -(F_a)^2 \quad (17.8)$$

EXAMPLE 52
A reduced eye of power +58.00 D accommodates 3.00 D. Its axial length is 22.25 mm. Calculate the near point distance. (*Figure 17.2*).

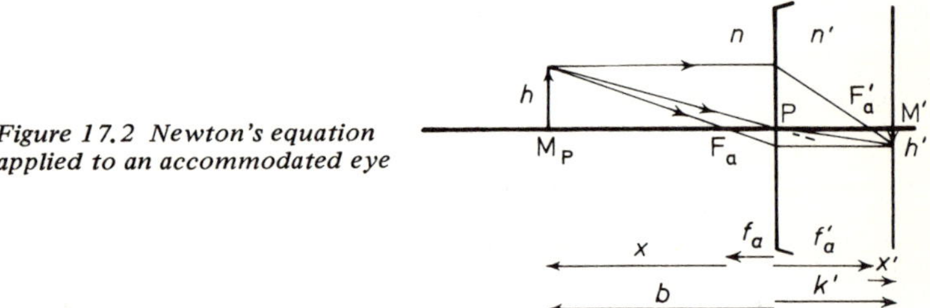


*Figure 17.2 Newton's equation applied to an accommodated eye*

*Solution 52*
*Step 1.* The power of the accommodated eye is (equation (12.1))

$$F_a = +58.00 + 3.00$$
$$= +61.00 \text{ D}$$

Hence, the second focal length is now (equation (12.7))

$$f'_a = \frac{1333}{+61.00}$$
$$= +21.85 \text{ mm}$$

*Step 2.* The distance $x'$ is given by equation (17.5); it may be written here as

$$x' = k' - f'_a$$

and so

$$x' = +22.25 - (+21.85)$$

$$= +0.40 \text{ mm}$$

Hence

$$X = \frac{1333}{+0.40}$$

$$= +3\,332.50 \text{ D}$$

*Step 3.* Applying equation (17.8), the near point vergence is

$$B = \frac{-(F_a)^2}{X'}$$

$$= \frac{-(+61.00)^2}{+3\,332.50}$$

$$= -1.12 \text{ D}$$

and so (equation (12.4))

$$b = \frac{100}{-1.12}$$

$$= -89.29 \text{ cm}$$

The error included in this solution is equal to the first focal length of the accommodated eye. This amounts to (equation (12.6))

$$f_a = \frac{-1000}{+61.00}$$

$$= -16.39 \text{ mm, or, say, } -1.64 \text{ cm}$$

## 17.4 Magnification

In the course of the derivation of Newton's equation it was shown that the magnification of an optical system can be described by

$$m = -\frac{x'}{f'}$$

which may be rewritten for the purposes of visual optics as

$$m = -\frac{x'}{f'_e}$$

This equation may be developed as follows:

$$\begin{aligned} m &= -\frac{k' - f'_e}{f'_e} \\ &= -\frac{k'}{f'_e} + \frac{f'_e}{f'_e} \\ &= -\frac{k'}{f'_e} + 1 \\ &= 1 - \frac{F_e}{K'} \\ &= 1 - \frac{F_e}{K + F_e} \end{aligned} \tag{17.9}$$

This expression has the advantage that the magnification caused by an eye, of an object situated in the far point plane, may be calculated on the basis of the ocular correction and the power of the eye, rather than the dioptric length.

EXAMPLE 53

A reduced eye of power +62.00 D has 3.00 D of myopia. An object of 5 mm is situated in the far point plane. Calculate the retinal image size.

*Solution 53*

*Step 1.* To find the magnification of the system, apply equation (17.9):

$$\begin{aligned} m &= 1 - \frac{+62.00}{-3.00 + 62.00} \\ &= 1 - \frac{+62.00}{+59.00} \\ &= 1 - (+1.050\,8) \\ &= -0.050\,8 \end{aligned}$$

*Step 2.* The retinal image size is given by equation (1.13):

$$h' = (-0.050\,8) \times 5$$
$$= -0.25 \text{ mm}$$

An alternative solution would be calculated as follows:
*Step 1.* Calculate the dioptric length of the eye (equation (5.2)):

$$K' = -3.00 + 62.00$$
$$= +59.00 \text{ D}$$

*Step 2.* The magnification of the system is given by equation (6.3):

$$m = \frac{-3.00}{+59.00}$$
$$= -0.050\,8$$

*Step 3.* This step is identical to Step 2 of the first solution.

In order to calculate the magnification of an accommodated eye, equation (17.9) may be modified to

$$m = 1 - \frac{F_e + A}{K + F_e + A}$$

or

$$m = 1 - \frac{F_a}{K + F_a} \qquad (17.10)$$

when the near object is imaged on the retina.

# Chapter Eighteen
# Ocular Rotation

## 18.1 Introduction

In normal binocular vision both eyes will be directed by the extra-ocular muscles towards the object of regard. When the visual axes of the eyes intersect at the object, a major objective of binocular vision has been achieved, namely the object is now imaged on the macula of each of the pair of eyes. If, for the purpose of bringing the visual axis to bear on the object, the eye has been rotated, it has done so around a certain point. In the reduced eye this point is known as the 'centre of projection'. It is indicated by the letter C and situated on the visual axis, 13 mm behind the principal point P (*Figure 4.5*). Details about the 'centre of rotation', that is, the point around which a real eye appears to rotate, are that it is situated 15 mm behind the anterior corneal surface and $1\frac{2}{3}$ mm nasally from the primary line (Emsley, 1955); further details were given by Fry (1962).

In calculations it is usually assumed that the eyes are originally in the primary position. This may be defined as the position taken up by a pair of eyes when looking at a distant object on the midline between the eyes, level with the eyes and with the head held erect.

## 18.2 Uncorrected vision with two eyes

When a pair of eyes is presented with an object placed on the midline between, level with and relatively near to the eyes, they will rotate from the primary position and their visual axes will converge. When the object is imaged on each macula, the visual axes will intersect at the object (*Figure 18.1*). The angle through which each eye has rotated from the primary position is represented by $\phi$. Since it is rather difficult to measure the actual distance between the centres of rotation or

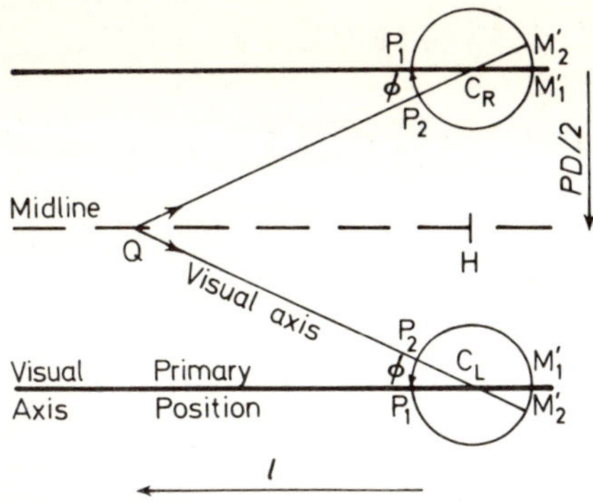


*Figure 18.1 The angle of convergence $\phi$*

projection of the eyes, it will be assumed that the interpupillary distance, commonly abbreviated as *PD*, represents the distance between the right eye's centre $C_R$ and that of the left eye, $C_L$.

The angle of convergence $\phi$ may be calculated thus:

$$\begin{aligned} \phi &= \angle P_1 C_R P_2 \qquad \text{(right eye)} \\ &= \angle P_1 C_L P_2 \qquad \text{(left eye)} \end{aligned}$$

In the diagram (*Figure 18.1*) the distance ½*PD* is measured below the axis, hence it is negative. Therefore, for the right eye

$$\tan \phi = -\frac{\tfrac{1}{2}PD}{l - P_1 C} \qquad (18.1)$$

NOTE

All equations will be derived for the right eye. The object distance $l$ is normally given with respect to the reduced surface P of an eye. However, the angle $\phi$ is measured at the centre of projection C which is situated 13 mm behind P, in the reduced eye. This difference must not be overlooked.

EXAMPLE 54

The *PD* of a pair of (reduced) eyes is 60 mm. They are presented with an object placed on their midline, 30 cm in front of the reduced surface. Calculate the angle of convergence through which each eye will have to rotate from the primary position (*Figure 18.1*).

*Solution 54*

Substitution into equation (18.1) of the data provided, shows

$$\tan\phi = -\frac{\tfrac{1}{2}(-60)}{-300-13}$$

$$= -0.095\,85$$

and so the angle of convergence $\phi = -5°\ 28'$.

In addition to the familiar measures of angle, others are used in connection with the eye. They are the 'prism dioptre' and the 'metre-angle'.

The angle whose tangent is 0.01 is 1 prism dioptre, usually written as 1Δ. The angle whose tangent is 0.02, subtends 2Δ, and so on. *See Figure 18.2.*

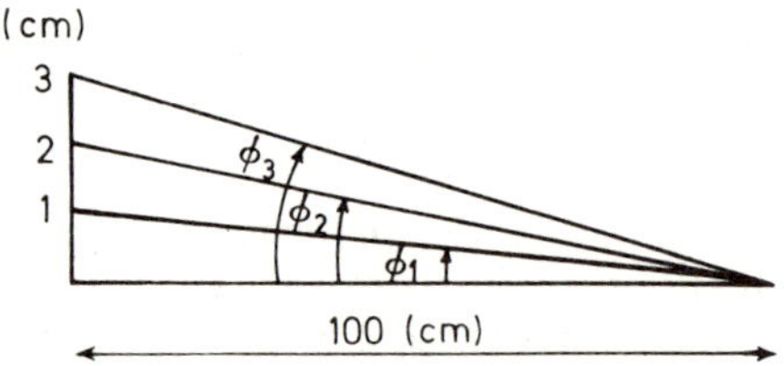


*Figure 18.2 The prism dioptre.* $\phi_1 = 1\Delta$; $\phi_2 = 2\Delta$; $\phi_3 = 3\Delta$

Returning to example 54, the angle of convergence $\phi = -9.585\Delta$, or about $-9.6\Delta$. Whether this pair of eyes is emmetropic or simply not corrected, would not alter the calculation of the theoretical angle of convergence. In near vision the object is usually situated below the level of the eyes. Consequently, the eyes need not only converge, but will also have to be rotated downwards. This latter movement is called 'depression'.

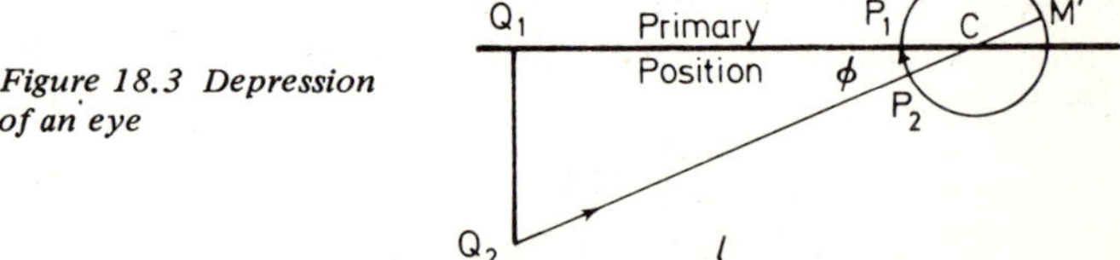


*Figure 18.3 Depression of an eye*

As will be seen from *Figure 18.3*, the calculation of the depression is basically the same as that of the convergence (compare with the upper

half of *Figure 18.1*). If the object $Q_2$ is situated 20 cm below the level of the eyes ($Q_1 Q_2 = -20$ cm) and the object distance were $-30$ cm, equation (18.1) could be employed with the proviso that the factor ½*PD* is replaced by the distance $Q_1 Q_2$. Thus:

$$\tan \phi = -\frac{Q_1 Q_2}{l - P_1 C} \tag{18.2}$$

Substituting the values given above, the angle $\phi$ is

$$\tan \phi = -\left(\frac{-200}{-300 - 13}\right)$$

$$= -0.638\,98$$

or the angle of depression $\phi = -32^\circ\ 35'$.
Alternatively, $\phi = -63.898\Delta$, or, say, $-63.9\Delta$.

NOTE
In both equation (18.1) and (18.2) all distances must be expressed in the same unit.

The above example forms also a convenient background for the introduction of another unit, namely the metre-angle (MA). This is a measure of the angle through which *one eye* rotates from the primary position to fixate a near object. It is, therefore, a measure of convergence such as represented by the angle $\phi$ in *Figure 18.1.*

When an eye has converged from the primary position to a near object situated on the midline between the eyes, 1 m in front of the centre of projection, it has converged 1 MA. When such an object is situated at ½ m, the eye will have to converge 2 MA. It will be clear that this unit is similar to the reciprocal metre used to measure the curvature of a surface.

In practice, the distance from the centre of projection to the near object on the midline, may be measured along the midline because the angle $\phi$ is usually small. In other words, instead of measuring the object distance $C_R Q$ or $C_L Q$, it may be substituted by the distance HQ (*Figure 18.1*). The difference is insignificant.

The number of metre-angles an eye has converged, may be calculated from the formula

$$\phi = \frac{1}{HQ}\ \text{MA} \tag{18.3}$$

or

$$\phi = \frac{1}{l - P_1 C}\ \text{MA} \tag{18.4}$$

where all distances are measured in metres.

This unit is very convenient with respect to accommodation in the emmetropic eye: when the eye looks at a near object at $\frac{1}{3}$ m, it will accommodate 3.00 D and converge 3.00 MA where the distance PC has been ignored.

The problem with this measure is, however, that the eye's uniocular *PD* has to be taken into account when converting metre-angles into prism dioptres or degrees. Therefore, 1 MA measured for different patients (or even for different eyes!) usually presents a slight variation in the number of prism dioptres and degrees. Consider *Figure 18.1.* The convergence of each eye (in prisms dioptres) is given by

$$\phi = -\frac{\frac{1}{2}PD \text{ (in centimetres)}}{\text{HQ (in metres)}} \Delta \qquad (18.5)$$

or

$$\phi = -(\text{number of MA}) \times (\tfrac{1}{2}PD \text{ in centimetres})\Delta \qquad (18.6)$$

Thus, for ½*PD* = 3.0 cm, 1 MA = 3.0Δ and 5 MA = 15.0Δ, but
for ½*PD* = 3.5 cm, 1 MA = 3.5Δ and 5 MA = 17.5Δ.

As a result of this complication, the metre-angle is not used in clinical practice, but is useful in concept.

## 18.3 Corrected ametropia

When the visual axis of an ametropic eye does not pass through the optical centre of the spectacle lens by which it is corrected, it will not receive an image of an object placed straight ahead of it, on its macula. This is due to the prismatic effect of the lens which may be defined as

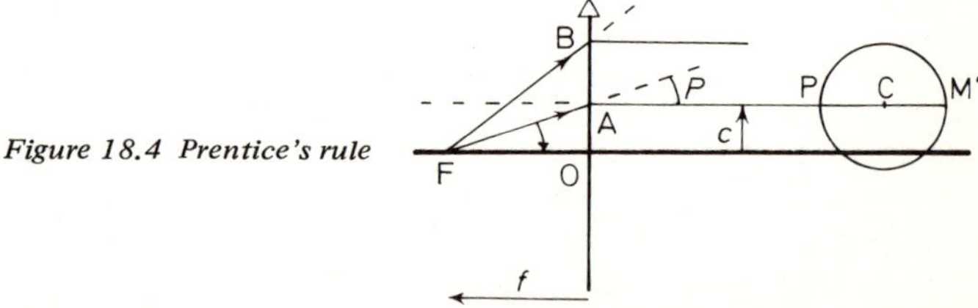


*Figure 18.4 Prentice's rule*

the deviation produced by a lens when a ray of light does not pass through its optical centre. Consider *Figure 18.4.* A point object is

placed at the first principal focus F of the optical axis FO of lens $F$. After refraction by this positive lens all rays originating from this point object will be parallel to the optical axis. It is clear that the ray passing through point B is subject to a greater deviation from its original course than the ray passing through point A. The deviation depends on the power of the lens and the distance from the optical centre at which the ray of light is incident on the lens, and it is usually measured in prism dioptres. Hence the angle of deviation $P$ is given by

$$\begin{aligned} P &= \frac{\mathrm{OA}}{\mathrm{OF}} \\ &= \frac{c}{-f} \\ &= \frac{c}{n/F} \end{aligned}$$

For a lens in air, $n = 1$ and so

$$\begin{aligned} P &= \frac{c}{1/F} \\ &= cF \end{aligned} \tag{18.7}$$

where $c$ is the distance from the optical centre of the lens to the point of incidence of the ray under consideration, measured in centimetres, and $F$ is the power of the lens in dioptres. Hence, the angle $P$ will be given in prism dioptres. Equation (18.7) is known as Prentice's rule, and it is valid for any ray of light within the paraxial region. The magnitude of the prismatic effect is given by $P = |cF|$ and this is understood when equation (18.7) is used.

Returning to the opening statement of this section, it will be clear that the eye shown in *Figure 18.4* would not see an object situated straight ahead along PA since that object would not be imaged on its macula. Only a distant object situated along the line AF (on the assumption that the lens $F$ represents the eye's distance correction) would be imaged on the macula. This situation presents itself in near vision when a distance correction is worn (*Figure 18.5*). The lenses are centred for distance vision, that is the distance between their optical centres S will (normally) be equal to the *PD*. Assuming the lenses to be thin and the auxiliary axes QS to fall within the paraxial region, the image Q′ of object Q will be formed along these axes. Their position may be calculated by means of the conjugate foci formula. However, the image position may also be constructed according to the rules given in section 1.1. For this purpose the perpendicular from the object point Q to the

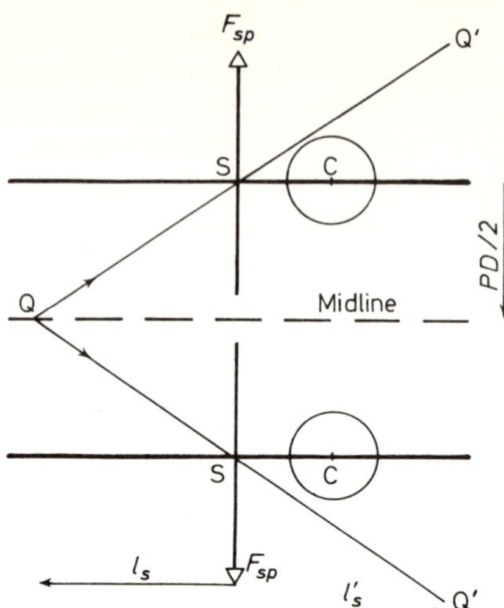


'igure 18.5 *The image* Q$'$ *formed by a distance correction of a near object* Q

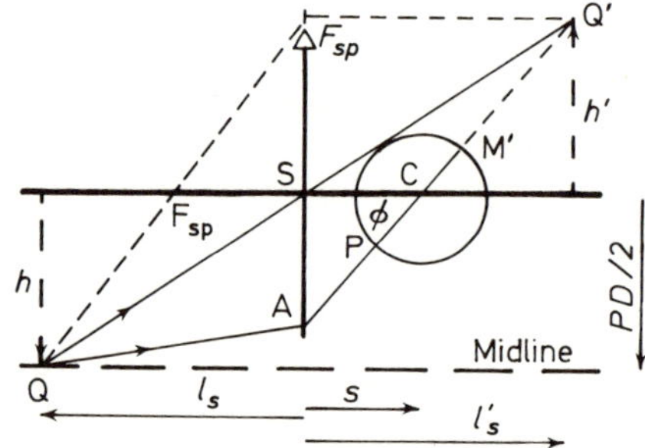


'igure 18.6 *To receive a foveal image of the near object* Q, *the eye must rotate* ɔ *that its visual axis* PCM$'$ *is directed towards the image* Q$'$ *formed by the spec-* ɪcle *lens of the object* Q. *The chief ray* QAPCM$'$ *gives rise to the foveal image*

ptical axis may be considered to represent the object size $h$ (*see* '*igure 18.6*). The image formed of this object by the lens $F_{sp}$, is indi- ated by its image size $h'$. In order to see the object Q clearly, that is, o receive an image of Q on its macula, the eye will have to rotate into ɪe direction of Q$'$ so that the macula M$'$, and Q$'$ are situated on the ame line. This line has then become an extension backwards of the ye's visual axis which lies along the line APCM$'$. From *Figure 18.7*

it can be seen that the ray QA is refracted by the lens $F_{sp}$, and continues along the visual axis of the eye. It will be understood that the eye will have to accommodate in order to see the object clearly. The construction of the prismatic effect of the lens at the point A is also

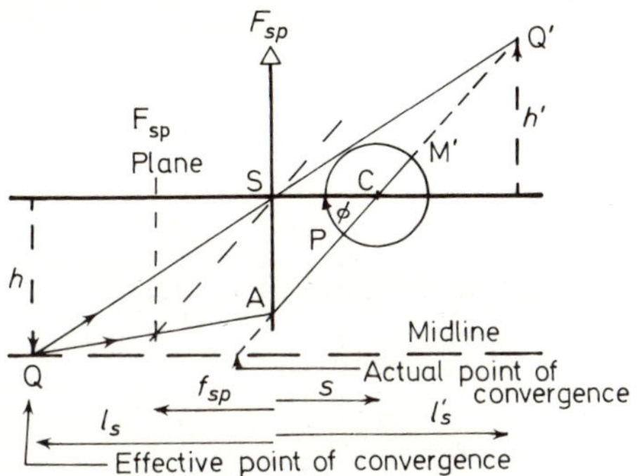


*Figure 18.7 The construction of the actual and effective point of convergence through a positive distance correction for a near object*

shown in *Figure 18.7* where an auxiliary axis through S is drawn parallel to the eye's visual axis. After refraction, the ray along the visual axis meets the auxiliary axis' ray in the first focal plane of the lens $F_{sp}$. This shows that, although the eye appears to look in the direction Q′A, it sees the object Q.

The calculation of the angle of convergence of the eye is shown in the following example.

EXAMPLE 55

A reduced eye is corrected for distance vision by a +10.00 DS lens at 12 mm. The centre of projection lies 13 mm behind its reduced surface. A point object is situated on the midline between the pair of eyes $\frac{1}{3}$ m in front of the spectacle plane. The ½*PD* is 30 mm. Calculate the angle through which the eye needs to rotate from the primary position so that the object will be imaged on its macula.

*Solution 55* (*see Figures 18.6* and *18.7*)

*Step 1.* The object vergence with respect to the spectacle lens, is given by equation (1.10):

$$L_s = \frac{1}{-(1/3)}$$

$$= -3.00 \text{ D}$$

*Step 2.* The image vergence with respect to the spectacle lens may be calculated from the conjugate foci formula:

$$L_s' = -3.00 + 10.00$$

$$= +7.00 \text{ D}$$

*Step 3.* The image distance from the spectacle lens is (equation (1.11)):

$$l_s' = \frac{100}{+7.00}$$

$$= +14.29 \text{ cm}$$

*Step 4.* To calculate the position of the image point Q′: using the magnification formula (equation (1.15)):

$$h' = \frac{L_s}{L_s'} \times h \qquad (\text{where } h = \tfrac{1}{2}PD)$$

$$= \frac{-3.00}{+7.00}(-30)$$

$$= +12.86 \text{ mm}$$

Alternatively, using similar triangles:

$$\frac{h'}{h} = \frac{l_s'}{l_s}$$

or

$$h' = \frac{hl_s'}{l_s}$$

$$= \frac{(-30)(+142.9)}{(-333.3)}$$

$$= +12.86 \text{ mm}$$

*Step 5.* The angle of rotation $\phi$ is given by

$$\tan\phi = -\frac{h'}{l_s' - s}$$

where $s$ = SC, that is ($d$ + PC).

Hence,

$$\tan\phi = -\frac{12.86}{142.9 - 25}$$

$$= -0.109\,1$$

Hence, angle $\phi = -6°14'$ or $-10.9\Delta$.

EXAMPLE 56

A reduced eye is corrected for distance vision by a −8.00 DS lens at 10 mm. The centre of projection is situated 15 mm behind the reduced surface. The eye is in the primary position. Calculate the amount of convergence required to obtain a macular view of an object situated 25 cm in front of the spectacle plane, on the midline between the eyes. ½$PD$ = 32 mm.

*Solution 56* (*see Figure 18.8*)

*Step 1.* Calculate the image distance from the spectacle lens:

$$\begin{array}{lcl} L_s = -4.00\text{ D} & \longleftarrow & l_s = -25.00\text{ cm} \\ F_{sp} = -8.00 & + & \\ \hline L_s' = -12.00 & \longrightarrow & l_s' = -8.33 \end{array}$$

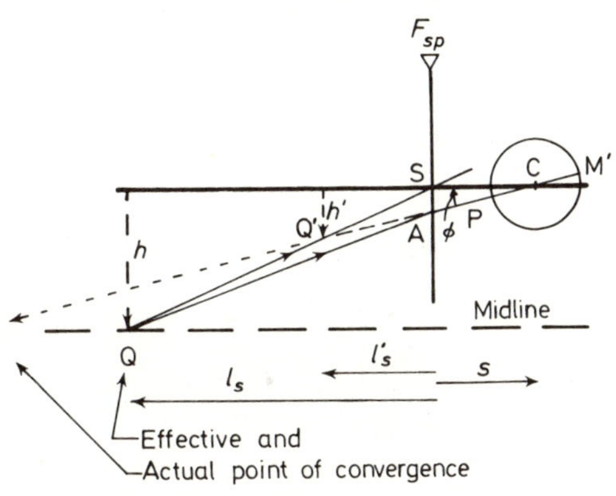


*Figure 18.8 Example 56; the actual and effective point of convergence for a negative distance correction*

*Step 2.* Calculate the position of the image with respect to the primary position of the eye:

$$h' = mh$$

$$= \frac{L_s}{L_s'} \times \frac{PD}{2}$$

$$= \frac{-4.00}{-12.00}(-32)$$

$$= -10.67\text{ mm}$$

Alternatively:

$$\frac{h'}{h} = \frac{l'_s}{l_s}$$

or

$$h' = \frac{hl'_s}{l_s}$$

$$= \frac{(-32)(-83.3)}{-250}$$

$$= -10.66 \text{ mm}$$

*Step 3.* Calculate the angle of rotation:

$$\tan\phi = -\frac{h'}{l'_s - s}$$

$$= -\frac{-10.67}{-83.3 - 25}$$

$$= -0.098\,5$$

Angle $\phi = -5°38'$ or about $-9.8\Delta$.

Occasionally a spectacle lens will be required to be decentred with respect to the interpupillary distance of the patient's eyes. This does not really change the approach shown above; it remains a matter of determining the position of the image produced by the spectacle lens of the object. This is, again, followed by the calculation of the angle through which the eye must rotate so that the object is eventually imaged on the macula. The next example deals with such a case.

EXAMPLE 57

An eye has a distance correction of +6.00 DS at 12 mm. This lens is decentred out 3 mm. An object point is situated on the midline, 36.36 cm in front of the spectacle plane. The ½*PD* is 30 mm, and the centre of projection of this (reduced) eye lies 13 mm behind the reduced surface. Determine the angle through which the eye must rotate from the primary position in order to see this object with the macula.

*Solution 57* (*see Figure 18.9*)
*Step 1.* Calculate the image position:

$$\begin{array}{rcl} L_s & = & -2.75\text{ D} \longleftarrow l_s = -36.36\text{ cm} \\ F_{sp} & = & +6.00 \\ \hline L_s' & = & +3.25 \longrightarrow l_s' = +30.77 \end{array} \quad +$$

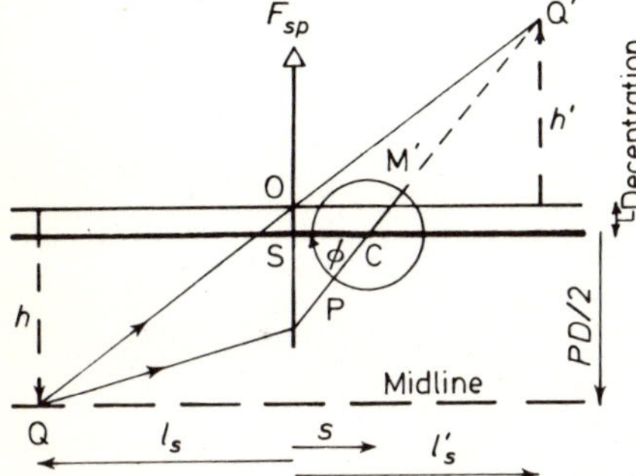


*Figure 18.9 Example 57*

*Step 2.* Calculate the position of the image with respect to the optical axis of the spectacle lens. Since the optical centre O of the spectacle lens $F_{sp}$ has been decentred out 3 mm (= SO), the distances $\boldsymbol{h}$ and $\boldsymbol{h'}$ must be measured with respect to the line parallel to the primary position line SC and passing through the point O, that is the optical axis of the spectacle lens. Now,

$$\begin{aligned} h &= \frac{1}{2}PD - \text{SO} \\ &= -30 - 3 \\ &= -33\text{ mm} \end{aligned}$$

The distance $h'$ may now be calculated:

$$\begin{aligned} h' &= mh \\ &= \frac{L_s}{L_s'} \times h \\ &= \frac{-2.75}{+3.25}(-33) \\ &= +27.92\text{ mm} \end{aligned}$$

Alternatively:

$$\frac{h'}{-h} = \frac{l'_s}{-l_s}$$

or

$$h' = \frac{-hl'_s}{-l_s}$$

$$= \frac{-33 \times 30.77}{-36.36}$$

$$= +27.93 \text{ mm}$$

*Step 3.* Calculate the angle of rotation. Since the angle of rotation $\phi$ is measured on the primary position line SC, the tangent of angle $\phi$ is given by

$$\tan \phi = -\frac{h' + SO}{l'_s - s}$$

$$= -\frac{27.92 + 3.00}{+309.2 - 25.00}$$

$$= \frac{-30.92}{+284.2}$$

$$= -0.108\,8$$

Hence, angle $\phi = -6°14'$ or about $-10.9\Delta$.

Although this far it has been assumed that the object is always situated on the midline, this is unlikely to be the case in everyday life. When an object is situated nearer to one eye than the other (*Figure 18.11*), unequal amounts of convergence will be required of the two eyes, and, in principle, unequal amounts of accommodation. This does not normally cause a problem. It will be appreciated that unequal object distances will give rise to unequal retinal image sizes assuming all other details to be the same. The small retinal image size difference causes no difficulty either and may be used in binocular vision as a clue in the perception of relative distances.

In *Figures 18.7* and *18.8* the effective and actual point of convergence have been indicated for situations involving a positive and a negative distance correction. If an eye has rotated sufficiently the effective point of convergence will coincide with the object point when the latter is situated on the midline. If the object point is not situated on the midline, the effective point of convergence is that point where the line AQ,

if necessary, after extension, crosses the midline. The actual point of convergence is the point where the visual axis M′CPA, when extended to the midline, crosses the midline (*Figures 18.7*, *18.8* and *18.10*). The difference in the position of the actual point of convergence with respect to the effective point of convergence, for a positive and a negative distance correction, is a function of the prismatic effect at the near visual point: prism base 'out' for a positive lens and prism base 'in' for a negative lens.

Finally, when a near object point is imaged by a low positive distance correction, the image may be formed at a distance further away then the object distance: object distance −¼ m, distance correction +2.00 D; image distance −½ m (*Figure 18.10*). Note that the image Q′ will be

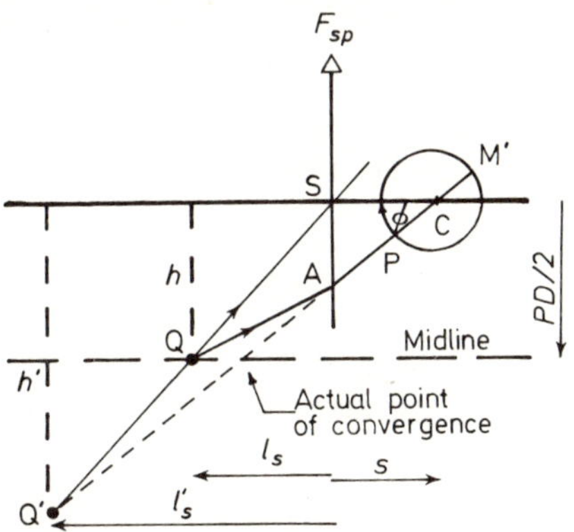


*Figure 18.10 The actual point of convergence for a low positive distance correction*

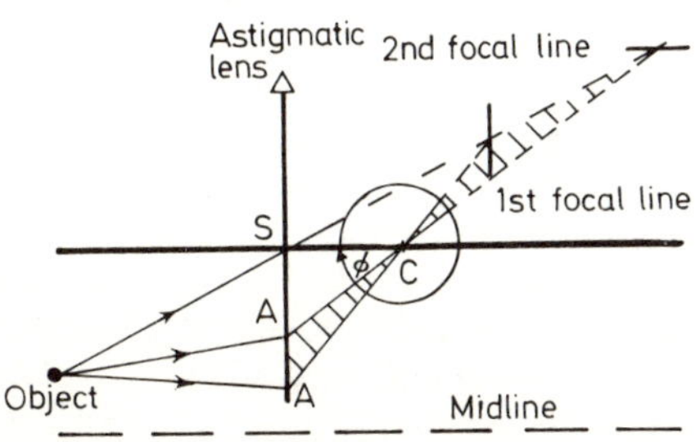


*Figure 18.11 The two near visual points* A *associated with the two focal lines formed by an astigmatic distance correction of a near point-shaped object*

situated on the auxiliary axis QS and *not* on the midline. When this situation is applied to a pair of eyes one may find that the image formed by the right-hand spectacle lens lies to the left of the midline,

and the image formed by the left-hand spectacle lens, lies to the right of the midline.

## 18.4 Astigmatism

When the correction of astigmatism was discussed (chapter 15), it was said that there are two far points and two near points, but that the principles of correction of this type of ametropia did not really differ from that of spherical ametropia. Problems associated with accommodation and near vision through a distance correction were discussed too (section 15.5).

In particular the question of which part of the interval of Sturm will be imaged on the retina, is relevant here. Since it appears that any point within this interval may be used by the ocular system (Fry, 1940; Obstfeld, 1976a), it becomes difficult to associate the angle $\phi$ with any specific point situated along the auxiliary axis through S (*Figure 18.11*). Hence, one could indicate the largest and smallest angle $\phi$, or the difference between the two (the hatched area in *Figure 18.11*), on the basis of the two focal lines formed of a point object by an astigmatic spectacle lens. Thence, the position of the two near visual points A associated with these extreme angles, can be calculated. In turn, the prismatic effect of the lens at these points and the actual point of convergence may be determined.

## 18.5 Anisometropia

This condition is marked by a considerable difference in the ametropia of each of a pair of eyes. A particular type of anisometropia is known as antimetropia. Here, one eye of the pair is myopic, the other hyperopic.

In uncorrected anisometropia the retinal image size will depend on the type of ametropia. If both eyes are refractively ametropic, their basic retinal image sizes will be the same. However, if the eyes are subject to axial ametropia, the longer eye will have a larger basic retinal image. If the difference in basic retinal image size is more than 5%, difficulties with the integration of the retinal images into a single binocular percept are likely to arise (*see* section 11.2).

In case of a pair of corrected anisometropic eyes, the spectacle lenses will produce spectacle magnification. In the case of refractive ametropia, the eye corrected by the more positive spectacle lens will have the larger retinal image. In case of axial ametropia, the eye corrected by the

more positive spectacle lens will have a smaller basic retinal image but it will be subject to spectacle 'magnification' ($SM > 1$); the eye corrected by the more negative spectacle lens will have a larger basic retinal image but it will be subject to spectacle 'minification' ($SM < 1$). It will be clear that each case will have to be considered on its merits because of the number of mutations and their dimensions.

Another cause of discomfort could be the unequal amounts of accommodation that will be required. As will be remembered, the corrected myopic eye needs to accommodate less than the corrected hyperopic eye, for the same object distance. Also, in a pair of corrected anisometropic myopic eyes, the more myopic eye of the pair will need to accommodate less than the fellow-eye. Furthermore, there will be a vertical prismatic difference in near vision when the near visual point is situated some distance below the optical centre of the spectacle lenses. For instance, in antimetropia, the myopic eye will be subject to prism base down and the hyperopic eye to prism base up. Similarly, the amounts of convergence will differ (asymmetrical convergence) though this tends to cause less discomfort (if any) than the vertical prismatic effect.

## 18.6 Aphakia

In binocular vision bilateral uncorrected aphakia presents problems owing to blurred retinal images. The amount of convergence would be the same as for any uncorrected pair of eyes (section 18.2).

When bilateral aphakia is corrected for distance vision, the amount of convergence may be calculated as for any other type of corrected ametropia. However, it should not be forgotten that aphakic eyes cannot accommodate so that their retinal images of near objects will be blurred. If both eyes are provided with an addition for the near vision distance at which the object of regard is situated, the retinal images will be sharp.

The majority of problem cases are found amongst patients with unilateral aphakia. Since aphakia is mainly a refractive type of ametropia, the original retinal image size of each of a pair of eyes will have been very similar provided the amounts of ametropia of each eye were similar. Consequently the retinal image size difference between the phakic and the aphakic eye is mainly due to spectacle magnification, and will amount to 15% or thereabouts. This retinal image size difference will cause, in principle, an unsurmountable perception problem for single binocular vision; convergence of the type discussed would be unequal (asymmetrical) but, in the main, possible.

An additional factor preventing fusion of the two retinal images in binocular vision might be the absence of accommodation. If the aphakic eye were corrected by a contact lens, the spectacle magnification would be much reduced and fusion for the purpose of single binocular vision might become possible (*see* section 20.6). An even better solution (Nordlohne, 1975) would be the fitting of an intra-ocular implant: a plastics lens fitted inside the eye by means of an operation, or even during an operation for the removal of a cataractous crystalline lens. Depending on the type of intra-ocular implant such lenses are fitted in front of, behind, or in the iris aperture. From an optical point of view the type fitted behind the iris would be best since it would almost completely replace the crystalline lens and its properties. There is, however, an important exception: accommodation is not restored.

Notwithstanding the problems outlined above, spectacle corrections have been found acceptable to a number of patients with unilateral aphakia (Chaston, 1975) who appear to be able to take advantage of the good retinal image in the corrected aphakic eye whilst suppressing the image of the phakic eye. This, however, can only be determined during a clinical trial.

## 18.7 Aniseikonia

This condition is present when there is an inequality in the size and/or shape of the two retinal images of a pair of eyes. Apart from an optical cause, aniseikonia (an = not; isos = equal; eikon = image (Greek)) may be due to the anatomy of the retina (differences in the two eyes in the distribution of the retinal receptors) or the neural pathways.

The 'optical' type of aniseikonia may occur in anisometropia while the 'anatomical' type is thought to be present when the ametropia of both eyes is very similar. Thus, aniseikonia may occur in a pair of emmetropic eyes.

The retinal image size difference between the two eyes, especially if over 5% (section 11.2), causes difficulties with fusion of the two retinal images into a single mental percept. Aniseikonia may be corrected by an iseikonic lens, i.e. a lens, or lens system, incorporating a prescribed magnification to correct the aniseikonia* as well as correcting the ametropia of the eye concerned. If the eye is emmetropic, the iseikonic lens will have to be afocal. A pair of spectacle lenses designed to give the same spectacle magnification notwithstanding differences in power,

*(BS 3521:1962)

is known as a pair of isogonal lenses, and would, for instance, avoid aniseikonia in anisometropia. In eyes with (corrected) astigmatism, differences in the shape of the two retinal images of a pair of eyes may occur too (section 16.3). In such cases the meridional difference in spectacle magnification may be corrected (Halass, 1959; Jalie, 1977).

# Chapter Nineteen
# Ocular Catoptrics and Catadioptrics

## 19.1 Introduction

In section 1.4 the optics of mirrors known as catoptrics was summarized. When an optical system consists of a combination of refracting and reflecting components it is called a catadioptric system. In the eye both catoptric and catadioptric image formations take place. These images were originally described by Purkinje (1823). Later, Sanson (1837) described their diagnostic use. Hence, they are known as Purkinje–Sanson images. They are made use of in keratometry (ophthalmo-)phakometry and to determine the position of the optical axis of the eye and associated angles.

## 19.2 Catoptrics: Purkinje–Sanson image I

The first image is formed by reflection of an object at the anterior surface of the cornea. This description is not completely accurate since the reflection takes place at the air-tears layer interface (*see* section 2.2.1). However, the tears layer may be ignored since it is very thin indeed, and may be considered to have concentric surfaces which do not materially affect the optical ray path.

The calculation of the details of the image is based on Gullstrand's No.2 Simplified Eye (*Table 2.1*), the details of which have been summarized in *Figure 19.1.* Furthermore, it is assumed that a 500 mm high object is situated 1 m in front of the corneal apex. The calculation, in step-along fashion, is shown in *Table 19.1* and should be used in conjunction with *Figures 19.1* and *19.2.* The latter shows an approximate ray trace to an arbitrary scale.

TABLE 19.1
Purkinje–Sanson image I, II, III and IV.
Object distance: 1 m; object size: 500 mm

| Step number* | Expression | Vergences (D) | | Distances (mm) | | Magnification | Image size (mm) | Position (mm) | Rel. Bright-ness |
|---|---|---|---|---|---|---|---|---|---|
| | | Sym-bol | Calculation | Sym-bol | Calculation | | | | |
| | | | | | *IMAGE I* | | | | |
| 1 | 1.10 | $L_1$ | $\frac{1}{-1.00} = -1.00$ | | | | | | |
| 2 | 1.35 | $F_1$ | $\frac{-2\times1000}{+7.70} = -259.74$ + | | | | | | |
| 3 | 1.30/1.15/ 1.13 | $L'_1$ | $-260.74 \rightarrow$ | $l'_1$ | $\frac{-1\times1000}{-260.74} = +3.84$ | $m_1 = \frac{-1.00}{-260.74} = +0.003\ 8$ | +1.90 | +3.84 | 1.00 |
| | | | | | *IMAGE II* | | | | |
| 1 | | $L_1$ | $-1.00$ | | | | | | |
| 4 | 1.4 | $F_1$ | $\frac{0.376\times1000}{+7.70} = +48.83$ | | | | | | |
| 5 | 1.12/1.11 1.15 | $L'_1$ | $+47.83 \rightarrow$ | $l'_1$ | $\frac{1.376\times1000}{+47.83} = +28.77$ | $m_1 = \frac{-1.00}{+47.83} = -0.020\ 9$ | | | |
| 6 | | | | $t_1$ | $+0.5$ − | | | | |
| 7 | 1.10 | $L_2$ | $\frac{1.376\times1000}{+28.27} = +48.67 \leftarrow$ | $l_2$ | $+28.27$ | | | | |
| 8 | 1.35 | $F_2$ | $\frac{-2(1.376)1000}{+6.8} = -404.71$ + | | | | | | |
| 9 | 1.30/1.15 1.13 | $L'_2$ | $-356.04 \rightarrow$ | $l'_2$ | $\frac{-1.376\times1000}{-356.04} = +3.86$ | $m_2 = \frac{+48.67}{-356.04} = -0.136\ 7$ | +1.43 | +4.36 | 0.01 |

| Step number* | Expression | Vergences (D) Symbol | Vergences (D) Calculation | Distances (mm) Symbol | Distances (mm) Calculation | Magnification | Image size (mm) | Position (mm) | Rel. Brightness |
|---|---|---|---|---|---|---|---|---|---|
| | | | | | *IMAGE III* | | | | |
| 7 | | $L_2$ | $+48.67$ | | | | | | |
| 10 | 1.4 | $F_2$ | $\frac{-0.04\times 1000}{+6.8} = -5.88$ + | | | | | | |
| 11 | 1.12/1.11/ 1.15 | $L_2$ | $+42.79$ | $\rightarrow l'_2$ | $\frac{1.336\times 1000}{+42.79} = +31.22$ | $m_2 = \frac{+48.67}{+42.79} = +1.137\,4$ | | | |
| 12 | | | | $t_2$ | $+3.1$ − | | | | |
| 13 | 1.10 | $L_3$ | $\frac{1.336\times 1000}{+28.12} = +47.51$ | $\leftarrow l_3$ | $+28.12$ | | | | |
| 14 | 1.35 | $F_3$ | $\frac{-2(1.336)1000}{+10.0} = -267.20$ + | | | | | | |
| 15 | 1.30/1.15/ 1.13 | $L'_3$ | $-219.69$ | $\rightarrow l'_3$ | $\frac{-1.336\times 1000}{-219.69} = +6.08$ | $m_3 = \frac{+47.51}{-219.69} = -0.216\,3$ | +2.57 | +9.68 | 0.008 |
| | | | | | *IMAGE IV* | | | | |
| 13 | | $L_3$ | $+47.51$ | | | | | | |
| 16 | 1.4 | $F_3$ | $\frac{0.077\times 1000}{+10.0} = +7.70$ + | | | | | | |
| 17 | 1.12/1.11/ 1.15 | $L'_3$ | $+55.21$ | $\rightarrow l'_3$ | $\frac{1.413\times 1000}{+55.21} = +25.59$ | $m_3 = \frac{+47.51}{+55.21} = +0.860\,5$ | | | |
| 18 | | | | $t_3$ | $+3.6$ − | | | | |
| 19 | 1.10 | $L_4$ | $\frac{1.413\times 1000}{+21.99} = +64.26$ | $\leftarrow l_4$ | $+21.99$ | | | | |
| 20 | 1.35 | $F_4$ | $\frac{-2(1.413)1000}{-6.0} = +471.00$ + | | | | | | |
| 21 | 1.30/1.15/ 1.13 | $L'_4$ | $+535.26$ | $\rightarrow l'_4$ | $\frac{-1.413\times 1000}{+535.26} = -2.64$ | $m_4 = \frac{+64.26}{+535.26} = +0.120\,1$ | −1.23 | +4.56 | 0.008 |

*NOTE: Step Numbers are in consecutive order. The first Step Number of Images II, III and IV refers to the same Step Number used earlier in the Table.

The Purkinje–Sanson image I is used in keratometry (measurement of the radius of curvature of the anterior surface of the cornea; keratometers are usually calibrated to provide the dioptric power of the cornea, too, as discussed in sections 2.2.1 and chapter 20), phakometry

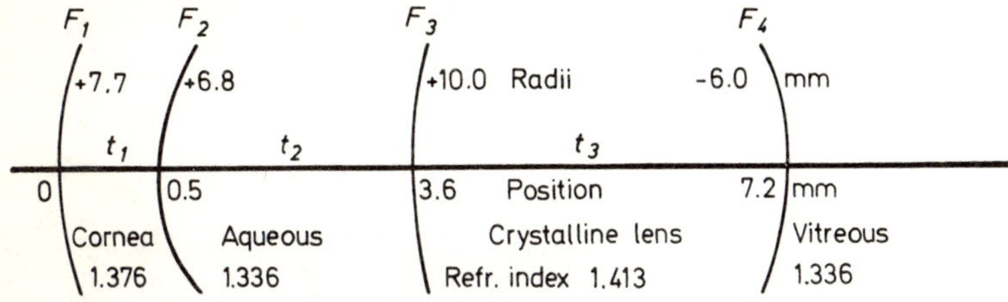


*Figure 19.1 Summary of the details of Gullstrand's no.2 eye used for the calculation of the Purkinje–Sanson images*

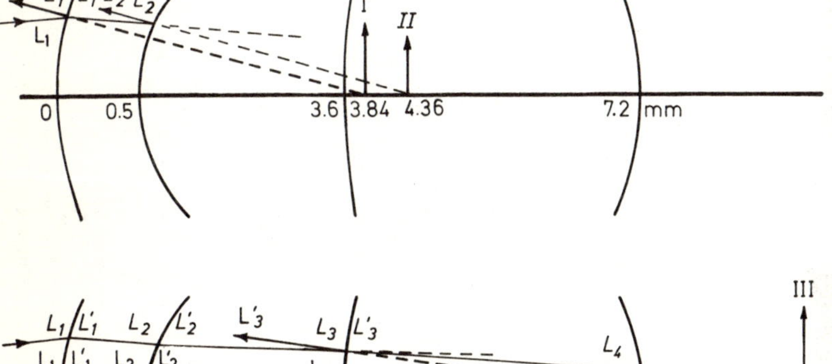


*Figure 19.2 The formation, relative sizes and positions of Purkinje–Sanson images I and II (above) and III and IV (below)*

(measurement of the radii of curvature of the crystalline lens whereby the Purkinje–Sanson image I is used as a comparison since the associated corneal radius can be measured directly with a keratometer) and ophthalmometry (measurement of the eye).

In *Table 19.1* the relative brightness of the image is taken as unity; the values for the other Purkinje–Sanson images are those given by Le Grand (1964). The reflection factor of the anterior corneal surface was discussed in section 2.2.4.

## 19.3 Catadioptrics: Purkinje–Sanson images II, III and IV

Purkinje–Sanson image II (*Figures 19.2* and *Table 19.1*) is formed by refraction at the anterior and reflection at the posterior corneal surface. As can be seen in *Figure 19.2*, Purkinje–Sanson images I and II are situated very close to each other and, when observed along the primary position line, image II would appear to be obscured by image I (but *see Table 19.2*). To make matters worse, image II is very dim indeed, owing to the very slight difference in refractive index between cornea and aqueous (*see also* section 2.2.4). Hence, Purkinje–Sanson image II is not normally observed. It may be observed though when the object is placed, say, 50° from the visual axis. The observation should then be made approximately along this axis.

Purkinje–Sanson image III is formed by refraction at both corneal surfaces and by reflection at the anterior lens surface. Purkinje–Sanson image IV is formed by refraction at both corneal surfaces as well as the anterior lens surface, and by reflection at the posterior lens surface.

Although they are dimmer than image II, image III can be seen because it is situated well beyond the other three images, and image IV, since it is inverted, can be recognized more easily.

Images III and IV are used in phakometry; since the size of the reflected images is proportional to the radius of the reflecting surface, the radius of the surface can be calculated when the object and image sizes as well as the refractive indices are known. Fincham (1935, 1937) investigated the changes taking place during accommodation, and was able to calculate the differences in the radii from photographs. Sorsby *et al.* (e.g. 1957 and 1962) investigated the normal distribution curves of the components of the eye using the above mentioned techniques.

The pencils of light giving rise to Purkinje–Sanson images III and IV are also partially reflected back into the eye by the anterior corneal surface. This gives rise to a further number of Purkinje–Sanson images.

Pipe (1975) pointed out that the images shown in *Figure 19.2* are, with the exception of image I, not those seen by an observer. These images act as 'objects'; the pencils emerging from the anterior corneal surface and originating from these 'objects', give rise to the images seen by the observer. The position of the latter, apparent images have been calculated in *Table 19.2*. When comparing the image 'position'

TABLE 19.2
Apparent Purkinje–Sanson images II, III and IV

| Step number* | Vergences (D) | | Distances (mm) | | Image position (mm) | Magnification | Image size (mm) |
|---|---|---|---|---|---|---|---|
| | Symbol | Calculation | Symbol | Calculation | | | |
| | | | *IMAGE II* | | | | |
| 9 | | | $l'_2$ | +3.86 | | | |
| 22 | | | $t_1$ | +0.5 + | | | |
| 23 | $L_3$ | $\frac{-1.376\times1000}{+4.36} = -315.60 \leftarrow$ | $l_3$ | +4.36 | | | |
| 24 | $F_1$ | $\frac{-0.376\times1000}{-(+7.70)} = +48.83$ + | | | | | |
| 25 | $L'_3$ | $-266.77 \rightarrow$ | $l'_3$ | $\frac{-1\times1000}{-266.77} =$ | +3.75 | $m_3 = \frac{-315.60}{-266.77} = +1.183\ 0$ | +1.70 |
| | | | *IMAGE III* | | | | |
| 15 | | | $l'_3$ | +6.08 | | | |
| 26 | | | $t_2$ | +3.1 + | | | |
| 27 | $L_4$ | $\frac{-1.336\times1000}{+9.18} = -145.53 \leftarrow$ | $l_4$ | +9.18 | | | |
| 28 | $F_2$ | $\frac{+0.04\times1000}{-(+6.80)} = -5.88$ + | | | | | |
| 29 | $L'_4$ | $-151.41 \rightarrow$ | $l'_4$ | $\frac{-1.376\times1000}{-151.41} = +9.09$ | | $m_4 = \frac{-145.53}{-151.41} = +0.961\ 2$ | |
| 30 | | | $t_1$ | +0.5 + | | | |
| 31 | $L_5$ | $\frac{-1.376\times1000}{+9.59} = -143.48 \leftarrow$ | $l_5$ | +9.59 | | | |
| 32 | $F_1$ | $\frac{-0.376\times1000}{-(+7.70)} = +48.83$ + | | | | | |
| | | | | −1×1000 | | −143.48 | |

| Step number* | Vergences (D) | | | Distances (mm) | | Image position (mm) | Magnification | Image size (mm) |
|---|---|---|---|---|---|---|---|---|
| | Symbol | Calculation | | Symbol | Calculation | | | |
| | | | | IMAGE IV | | | | |
| 21 | | | | $l'_4$ | −2.64 | | | |
| 34 | | | | $t_3$ | +3.6 + | | | |
| 35 | $L_5$ | $\frac{-1.413\times 1000}{+0.96}$ | = −1474.87 ← | $l_5$ | +0.96 | | | |
| 36 | $F_3$ | $\frac{-0.077\times 1000}{-(+10.0)}$ | = +7.70 + | | | | | |
| 37 | $L'_5$ | | −1464.17 → | $l'_5$ | $\frac{-1.336\times 1000}{-1464.17}$ = +0.91 | | $m_5 = \frac{-1471.87}{-1464.17} = +1.005\ 3$ | |
| 38 | | | | $t_2$ | +3.1 + | | | |
| 39 | $L_6$ | $\frac{-1.336\times 1000}{+4.01}$ | = −333.17 ← | $l_6$ | +4.01 | | | |
| 40 | $F_2$ | $\frac{+0.04\times 1000}{-(+6.80)}$ | = −5.88 + | | | | | |
| 41 | $L'_6$ | | −339.05 → | $l'_6$ | $\frac{-1.376\times 1000}{-339.05}$ = +4.06 | | $m_6 = \frac{-333.17}{-339.05} = +0.982\ 7$ | |
| 42 | | | | $t_1$ | +0.5 + | | | |
| 43 | $L_7$ | $\frac{-1.376\times 1000}{+4.56}$ | = −301.75 ← | $l_7$ | +4.56 | | | |
| 44 | $F_1$ | $\frac{-0.376\times 1000}{-(+7.70)}$ | = +48.83 + | | | | | |
| 45 | $L'_7$ | | −252.92 → | $l'_7$ | $\frac{-1\times 1000}{-252.92}$ = | +3.95 | $m_7 = \frac{-301.75}{-252.92} = +1.1931$ | −1.45 |

*NOTE: The first Step Number of each image refers to the last Step Number of the same image in *Table 19.1*. Thence, consecutive numbers are used.

in *Table 19.1* and the 'image distance' in *Table 19.2* as well as the image sizes of each of the Purkinje–Sanson images, they will be seen to be very similar. This may well be the reason why this point is not usually made.

## 19.4 Axes and Angles

A number of axes are recognized in the eye. The Purkinje–Sanson images are used to determine the position of the optical (or optic) axis. When the images are observed they do not usually appear to be in line. However, when the eye is slowly rotated, the images may come in line with each other to a greater or lesser degree. Since it appears almost impossible to produce a perfect alignment of the images in the human eye, it follows that the optical components do not form a symmetrical optical system. The optical axis (ONO′) is taken to pass

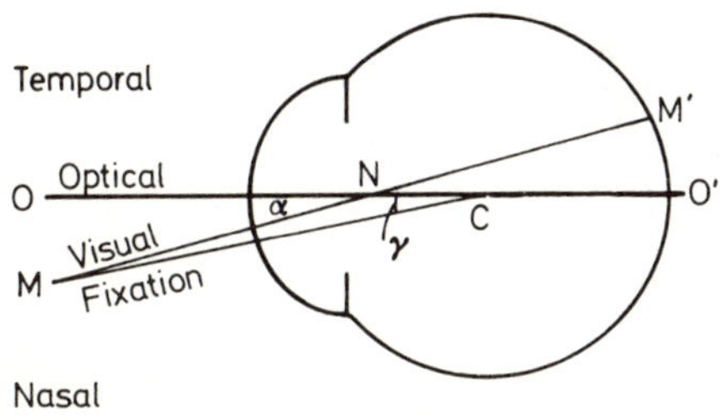


*Figure 19.3 The angles alpha and gamma*

through the optical centres of the eye's optical components (*Figures 19.3* and *4.3*) and the observer's eye when the best line-up of the Purkinje–Sanson images has been achieved.

The visual axis or line is the imaginary line passing through the object of regard M, the nodal point N of the eye and the macula M′ (*Figures 19.3* and *4.3*). This axis varies slightly from the fixation axis which passes through the object M and the centre of rotation C (*Figure 19.3*).

The pupillary axis is the imaginary normal to the cornea passing through the pupil's centre (*Figure 19.4*).

The following angles may be determined (all lie in the horizontal plane and are shown for the right eye):

angle alpha ($\alpha$): Between optical and visual axis (*Figures 19.3* and *4.3*); usually subtending about 5°.

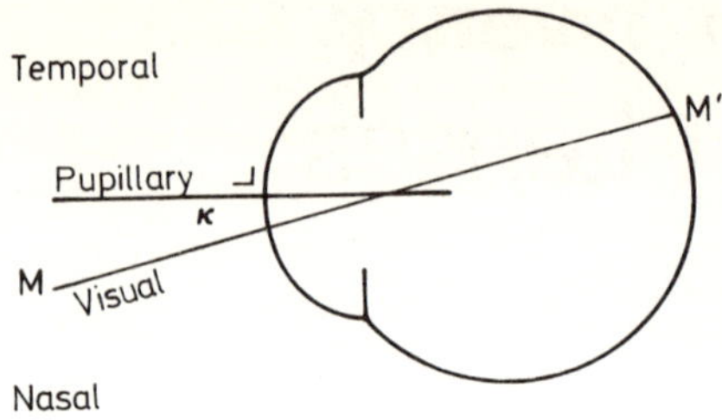


*Figure 19.4 The angle kappa*

angle gamma ($\gamma$): Between optical and fixation axis (*Figure 19.3*); for a distant object angles alpha and gamma subtend the same angle, but for a near object angle gamma is a little smaller than angle alpha.

angle lambda ($\lambda$): Between pupillary axis and the 'line of sight' which is parallel to the visual axis (for distant objects) and passes through the centre of the entrance pupil (Michaels, 1975; but see Le Grand, 1964). This angle is usually called:

angle kappa ($\kappa$): *See Figure 19.4.*

The line of sight is usually situated nasally to the pupillary axis; it is then denoted as 'positive'. When the line of sight is situated to the temporal side of the pupillary axis, the angle kappa is denoted 'negative'.

NOTE

Mention should be made of the visual angle which may be defined as the angle subtended by an object at the anterior nodal point of an eye (or the nodal point of the reduced eye).

# Chapter Twenty
# Contact Lenses

**The revision of the British Standard Specification dealing with terms relating to contact lenses has been under way for some time now. At the time of writing no revised version of BS 3521:1962 had been published. A draft revision dated December 1971 was sent to interested parties for comment. Attention has been paid to changes proposed in this draft, and some of the terms used in this chapter have been taken from it.**

## 20.1 Introduction

A discussion of a spectacle lens correction usually involves two optical components, namely the spectacle lens (which, in most cases, may be considered to be a 'thin' lens) and the eye. A discussion of a contact lens correction is rather more complicated since it involves a thick lens system (the contact lens), followed by a liquid lens, the anterior surface of the cornea, and components of the eye.

Certain terms used in the last sentence must be defined first. According to BS 3521:1962, they are:

contact lens: A lens worn in contact with the eyeball. It is generally divided into optic and haptic portions, but the latter may be absent.

Hence, definitions of:

haptic portion In general, that portion of a contact lens which lies in front of the sclera.

optic portion: In general, that part of a contact lens which lies in front of the cornea.

Furthermore:

liquid lens: The lens formed by the liquid lying between the cornea and the back surface of the optic portion.

There are two basic types of contact lens, namely:

1. The scleral lens: A contact lens with a haptic portion.
2. The corneal lens: A contact lens without a haptic portion.

These days, contact lenses are made of hard or soft plastics materials. They have different properties, both clinically and physically. This text will consider hard contact lenses only. Some optical problems peculiar to soft contact lenses are a matter of controversy at present. They have been discussed by Ford (1976), Bennett (1976) and Bier and Lowther (1977) in whose texts further references will be found.

The following discussion will hold good for both scleral and corneal contact lenses although there are differences owing to differences in the fitting philosophy. Furthermore, only the optic portion of any contact lens will be considered.

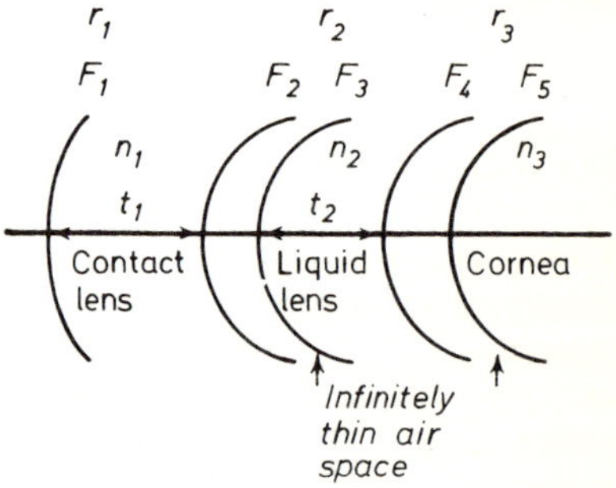

*Figure 20.1 The components of a contact lens system*

Returning to the theme of the first paragraph of this section, a contact lens may be considered for the purpose of calculations as consisting of four centred single spherical (or toroidal) refracting surfaces (*Figure 20.1*). The components will be denoted as follows:

the contact lens:

front radius: FOR (front optic radius) = $r_1$ (mm)
back radius: BOR (back optic radius) = $r_2$ (mm)
front surface power: $F_1$ (D)
back surface power: $F_2$ (D)
refractive index: $n_1$
thickness: $t_1$ (mm)

separated by an infinitely thin air space from the liquid lens:

front radius: $r_2$ (mm)
back radius: $r_3$ (mm) = 'K'-reading (mm) (*see* p.262)

| | |
|---|---|
| front surface power: | $F_3$ (D) |
| back surface power: | $F_4$ (D) |
| refractive index: | $n_2$ |
| thickness: | $t_2$ (mm) |

separated by an infinitely thin air space from the front surface of the cornea:

| | |
|---|---|
| front radius: | $r_3$ (mm) = 'K'-reading (mm) |
| front surface power: | $F_5$ (D) |
| refractive index: | $n_3$ |

The radius of curvature of the front surface of the cornea is measured with an instrument called a keratometer (also known as ophthalmometer), hence the term 'K'-reading.

NOTE
The 'K'-reading is in no way related to the ocular correction $K$.

## 20.2 Optics of contact lenses

The relationship between the front surface of the cornea and the back surface of the contact lens will depend on the type of contact lens used and the fitting philosophy adopted by the contact lens practitioner. These two surfaces are the limiting surfaces of the liquid lens. Owing to differences in refractive index (*see* chapter 2: contact lens $n_1 = 1.49$; liquid lens $n_2 = 1.336$; cornea $n_3 = 1.376$), these surfaces have different powers (in air), namely

$$F_1 = \frac{n_1 - n_{air}}{r_1} = \frac{(1.49 - 1)1000}{r_1} = \frac{490}{r_1} \quad (20.1)$$

$$F_2 = \frac{n_{air} - n_1}{r_2} = \frac{(1 - 1.49)1000}{r_2} = \frac{-490}{r_2} \quad (20.2)$$

$$F_3 = \frac{n_2 - n_{air}}{r_2} = \frac{(1.336 - 1)1000}{r_2} = \frac{336}{r_2} \quad (20.3)$$

$$F_4 = \frac{n_{air} - n_2}{r_3} = \frac{(1 - 1.336)1000}{r_3} = \frac{-336}{r_3} \quad (20.4)$$

$$F_5 = \frac{n_3 - n_{air}}{r_4} = \frac{(1.376 - 1)1000}{r_4} = \frac{376}{r_4} \quad (20.5)$$

The thickness of the contact lens is arrived at in the same way as is the thickness of a spectacle lens. Basic details may be found in any textbook on ophthalmic lenses (Bennett, 1968b; Jalie, 1977).

The effect of the thickness on the power of a contact lens cannot be overlooked. The reason is that the surface powers of contact lenses are very high. Consider the following example.

EXAMPLE 58

A contact lens is held in air. It has a centre thickness of 1 mm and a front surface radius of +8.00 mm. A pencil of light from a distant point object is incident on the front surface of this lens. Calculate the vergence of the pencil incident on the back surface.

*Solution 58*

*See Figure 20.2.*

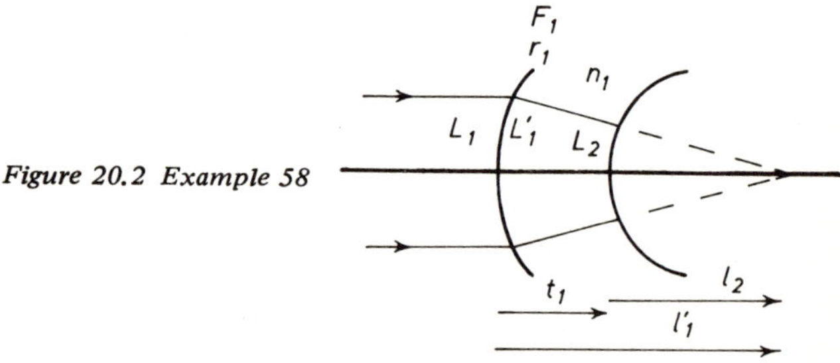


*Figure 20.2 Example 58*

Using the step-along method, we find:

$$L_1 = 0.00\text{ D}$$

$$F_1 = \frac{n_1 - n_{air}}{r_1} = \frac{490}{+8.00} = +61.25 \quad +$$

$$L'_1 = +61.25 \longrightarrow l'_1 = +16.33\text{ mm}$$

$$t_1/n_1 = \frac{1}{1.49} = +0.67 \quad -$$

$$L_2 = +63.86 \longleftarrow l_2 = +15.66$$

which shows that in this case a 1 mm centre thickness changes the vergence by more than 2.5 D.

NOTE

In calculations involving contact lenses, distances are most conveniently measured in millimetres.

Although the thickness of the liquid lens associated with a corneal lens is usually assumed to be negligible, it cannot always be ignored. This applies even more to the liquid lens associated with most scleral lenses since these can be quite 'thick'. These factors will be considered in more detail below.

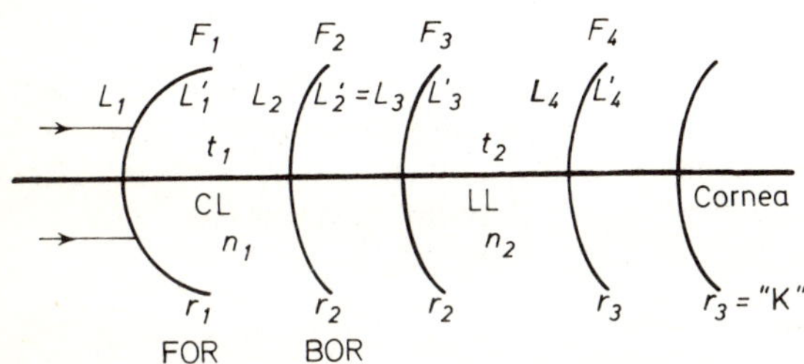


*Figure 20.3 Example 59; step-along method.* CL = *contact lens;* LL = *liquid lens*

The following example sets out a complete calculation through a contact lens correction starting with the spectacle correction and employing the step-along method throughout. The outcome is the front optic radius of the corneal lens (*Figure 20.3*).

EXAMPLE 59

| Details: | spectacle correction | +6.00 DS |
|---|---|---|
| | vertex distance | 12 mm |
| | 'K'-reading | 8.20 mm |
| | liquid lens thickness | 0.05 mm |
| | BOR | 8.10 mm |
| | contact lens thickness | 0.5 mm |

Calculate the front optic radius of the contact lens.

*Solution 59*

*Step 1.* Calculate the ocular correction.

$$F_{sp} = +6.00\text{ D} \longrightarrow f'_{sp} = +16.67\text{ cm}$$
$$d = +1.2 \quad -$$
$$K = +6.46 \longleftarrow k = +15.47$$

*Step 2.* The back vertex power (*BVP*) of the contact lens–liquid lens system should thus provide +6.46 D (= *K*) at the cornea. Calculating

from the cornea towards the front surface of the contact lens, opposite to the direction of the incident light, we obtain (*Figure 20.3*):

$$
\begin{array}{llllllll}
L'_4 & = BVP & = & & +6.46\ \text{D} & & & \\
F_4 & = \dfrac{-336}{r_3} & = \dfrac{-336}{+8.20} & = & \underline{-40.98}\ - & \text{(equation (20.4))} & & \\
L_4 & = & & & +47.44 \longrightarrow & l_4 = & & +21.08\ \text{mm} \\
 & & & & & t_2/n_2 = \dfrac{+0.05}{1.336} & = & \underline{+0.04}\ + \\
L'_3 & = & & & +47.35 \longleftarrow & l'_3 = & & +21.12 \\
F_3 & = \dfrac{336}{r_2} & = \dfrac{336}{+8.10} & = & \underline{+41.48}\ - & \text{(equation (20.3))} & & \\
L_3 & = L'_2 & = & & +5.87 & & & \\
F_2 & = \dfrac{-490}{r_2} & = \dfrac{-490}{+8.10} & = & \underline{-60.49}\ - & \text{(equation (20.2))} & & \\
L_2 & = & & & +66.36 \longrightarrow & l_2 = & & +15.07 \\
 & & & & & t_1/n_1 = \dfrac{+0.50}{1.49} & = & \underline{+0.34}\ + \\
L'_1 & = & & & +64.89 \longleftarrow & l'_1 = & & +15.41 \\
L_1 & = & & & \underline{0.00}\ - & & & \\
F_1 & = & & & +64.89 & & &
\end{array}
$$

*Step 3.* Equation (20.1) provides the front optic radius (FOR), thus:

$$r_1 = \frac{490}{F_1} = \frac{490}{+64.89} = +7.55\ \text{mm}$$

This calculation can be shortened by using equation (1.4) to determine the power of $(F_3 + F_2)$, namely

$$\frac{(1.336 - 1.490)}{+8.10} = -19.01\ \text{D}$$

In the course of the fitting of a scleral contact lens it is sometimes necessary to change the BOR because the relationship between the contact lens back surface and the front surface of the cornea is not optimum. The result of such a change is shown in the following example.

EXAMPLE 60

The BOR of a sighted scleral contact lens (that is, a lens having both front and back surfaces worked to give a certain back vertex power) is 8.00 mm. The lens is altered to make its BOR 7.95 mm, in order to

improve the fit. Calculate any alteration required in the final lens as compared with the original lens (*Figure 20.1*).

*Solution 60*

A change in the BOR affects the front surface of the liquid lens, and hence the power of this surface. The original front surface power in air was (equation (20.3))

$$F_3 = \frac{336}{+8.00} = +42.00 \text{ D}$$

The new front surface power will be

$$F_3 = \frac{336}{+7.95} = +42.26 \text{ D}$$

Hence the increase in the liquid lens power is +0.26 D. This should be counteracted by an alteration of the back vertex power of the contact lens of −0.26 D. However, the back surface power of the contact lens is affected too. The original back surface power in air was (equation (20.2))

$$F_2 = \frac{-490}{+8.00} = -61.25 \text{ D}$$

The new back surface power will be

$$F_2 = \frac{-490}{+7.95} = -61.64 \text{ D}$$

which is a change of −0.39 D. Thus, the contact lens–liquid lens system has undergone a power change of +0.26 + (−0.39) = −0.13 D. Only the front surface of the contact lens can be altered to counteract this overall change. When its power has been increased by +0.13 D the original back vertex power of the contact lens–liquid lens system will have been restored.

A rule-of-thumb based on this calculation, says:

*Rule 1*: for each 0.05 mm reduction (steepening) in the BOR, +0.12 D must be added to the front surface power of the contact lens to maintain the original back vertex power of the contact lens–liquid lens system.

When corneal lenses are fitted, it is impractical to alter the BOR of an exiting lens, as can be done with a scleral lens. Hence this must be considered in a different way, as follows.

EXAMPLE 61

The BOR of a corneal lens is 8.00 mm and its back vertex power is −3.00 D. A new lens of BOR 7.95 mm is to be supplied to improve the fit. Calculate the alteration to the *BVP* required for the new lens.

*Solution 61*

As before, the back vertex power of the contact lens–liquid lens system must remain the same. The change of BOR affects the front surface power of the liquid lens and, as shown in example 60, it is altered from +42.00 to +42.26 D. Hence, the back vertex power of the contact lens must be altered to counteract the change in the power of the liquid lens. Thus, the new contact lens must have a back vertex power of −3.00 − (+0.26) = −3.26 D.

Referring to example 60, it will be seen that this is the same change as that of the scleral lens where −0.39 D was added to the back surface and +0.13 D to the front surface. However, with corneal contact lenses one is not concerned with alterations to be made to existing surface powers (since this is impractical) and so a new contact lens having a different back vertex power from the original lens must be ordered.

The above leads to the next rule-of-thumb:

*Rule 2*: For each 0.05 mm reduction (steepening) of the BOR of a corneal lens, −0.25 D must be added to the *BVP* of the contact lens to maintain the original *BVP* of the contact lens–liquid lens system.

Another alteration encountered in clinical practice is that a scleral lens may need to be made thinner. This, too, has an effect on the back vertex power as shown in the following example.

EXAMPLE 62

Assume the front surface power of a contact lens to be $F_1 = +60.00$ D and the original lens thickness to be $t_1 = 0.60$ mm. The lens needs to be made thinner by 0.10 mm. Calculate the difference in back vertex power.

*Solution 62*

A change in the thickness does not affect the back surface power but only the effectivity of the front surface power at the back surface. Therefore, the change in the back vertex power is equal to the change in the effectivity of the front surface power at the back surface. This may be calculated from the effectivity formula (*see* section 9.2.2.2).

In the original lens the effectivity of the front surface power at the back surface for $t_1 = 0.60$ mm, was (*Figure 20.2*)

$$L_2 = \frac{F_1}{1-(t_1/n_1)F_1}$$

$$= \frac{+60.00}{1-(0.000\,6/1.490)(+60.00)}$$

$$= \frac{+60.00}{1-0.024\,2}$$

$$= +61.49 \text{ D}$$

The effectivity of the same front surface power at the new thickness of $t_1 = 0.50$ mm, is

$$L_2 = \frac{+60.00}{1-(0.000\,5/1.490)(+60.00)}$$

$$= \frac{+60.00}{1-0.020\,1}$$

$$= +61.23 \text{ D}$$

Hence, a change in the lens thickness of 0.10 mm produces a change in the back vertex power of $+61.23-(+61.49) = -0.26$ D.

A rule-of-thumb based on this calculation, says:

*Rule 3*: For each 0.1 mm reduction in the thickness of an existing contact lens, +0.25 D must be added to the back vertex power to maintain the original back vertex power.

NOTE

As can be shown (Neumueller, 1968) this rule applies only to a certain range of contact lens front surface powers, but is acceptable as a rule-of-thumb.

Another change encountered mainly during the fitting of scleral lenses, is a reduction in the liquid lens thickness owing to 'settling of the lens'. Since this affects the back vertex power of the contact lens–liquid lens system, the effectivity of the system at the cornea is given by (*see* section 9.2.2.2 and *Figure 20.3*)

$$L_4' = \frac{BVP_{CL}+F_3}{1-(t_2/n_2)(BVP_{CL}+F_3)} \tag{20.6}$$

(where $BVP_{CL}$ = back vertex power of the contact lens) as will be seen from the discussion of the change in the thickness of a contact lens, given above.

For the sake of simplicity, we will assume in the next example that the eye is fitted with an afocal lens. Hence, equation (20.6) reduces to

$$L'_4 = \frac{F_3}{1-(t_2/n_2)F_3} \tag{20.7}$$

EXAMPLE 63

Calculate the change in effectivity of a contact lens–liquid lens system when the liquid lens is reduced from 0.2 mm to 0.1 mm. The BOR of the afocal lens is 8.00 mm.

*Solution 63*

The back vertex power of the contact lens–liquid lens system is given by equation (20.7). The front surface power of the liquid lens may be calculated from equation (20.3):

$$\begin{aligned} F_3 &= \frac{336}{+8.00} \\ &= +42.00\text{ D} \end{aligned}$$

By substitution into equation (20.7) we find for the original thickness $t_2$ = 0.2 mm a back vertex power of

$$\begin{aligned} L'_4 &= \frac{+42.00}{1-(0.000\,2/1.336)(+42.00)} \\ &= \frac{+42.00}{1-0.006\,3} \\ &= \frac{+42.00}{0.993\,7} \\ &= +42.27\text{ D} \end{aligned}$$

and for the new thickness of $t_2$ = 0.1 mm a back vertex power of

$$\begin{aligned} L'_4 &= \frac{+42.00}{1-(0.000\,1/1.336)(+42.00)} \\ &= \frac{+42.00}{1-0.003\,1} \\ &= \frac{+42.00}{0.996\,9} \\ &= +42.13\text{ D} \end{aligned}$$

A decrease in the liquid lens thickness of 0.1 mm has thus effected a change in the back vertex power of the contact lens–liquid lens system of +42.13 − (+42.27) = −0.14 D.

As a rule-of-thumb we may say:

*Rule 4*: For each 0.1 mm reduction in the liquid lens thickness, +0.12 D must be added to the back vertex power of the contact lens–liquid lens system to maintain the original back vertex power.

In the case of a scleral lens, on reducing the thickness of the optic portion only (by altering the back surface), the thickness of the liquid lens is increased. Consequently, Rules 3 and 4 must be applied, namely per 0.1 mm:

| | |
|---|---|
| to compensate for the reduction of the thickness of the optic portion, add | +0.25 D |
| to compensate for the increase in the thickness of the liquid lens, add | −0.12 D + |
| add to the back vertex power of the contact lens | +0.12 D |

This addition must be made by a change in the front surface power of the contact lens.

Other details about these Rules have been given by Swaine (1956).

## 20.3 The pupil

### 20.3.1 INTRODUCTION

So far no account has been taken of the apparent position of the pupil of the real eye. In the reduced eye it is assumed that the apparent position of the pupil coincides with the reduced surface which is situated $1\frac{2}{3}$ mm behind the vertex of the real cornea (*see* section 4.2). In the optics of contact lenses it may be necessary to be more exact than that. The calculation of the position of the entrance pupil (En P) and exit pupil (Ex P) will be shown in the following section; the implications will be discussed in section 20.6.

### 20.3.2 THE ENTRANCE PUPIL

The entrance pupil is defined as the image of the aperture stop formed by that part of an optical system on the object side of the stop. For

our purpose this may be adapted to read: the entrance pupil is the image of the actual pupil formed by rays refracted by the cornea.

Only rays passing through the entrance pupil contribute to the retinal image formation. To calculate the position and size of the entrance pupil the data of Gullstrand's Schematic Eye No.1 will be used (*see Table 2.1* and *Figure 20.4*).

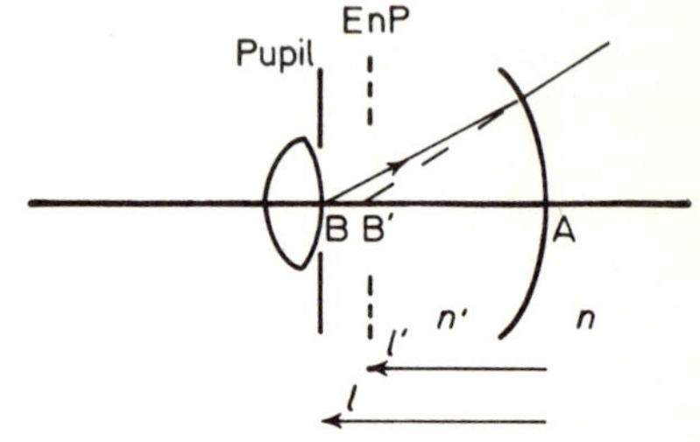


*Figure 20.4 The position of the entrance pupil* En P

If we assume that the plane of the real pupil coincides with the anterior pole B of the crystalline lens, then the object distance $l$ = BA (where A is the apex of the cornea). According to *Table 2.1*, the anterior lens surface is situated 3.6 mm behind the anterior corneal surface. Thus we have

$$L = \frac{n'}{l} = \frac{133.6}{-3.6} = -371.11 \text{ D} \leftarrow l = \quad -3.6 \text{ mm}$$

$$\begin{aligned} \text{corneal power } F &= \quad \frac{+43.05}{-328.06} + \\ L' &= \quad \phantom{x} \rightarrow l' = \frac{1000}{-328.06} = -3.05 \end{aligned}$$

This is not at all in agreement with the value of $1\frac{2}{3}$ mm used in the reduced eye.

The size of the entrance pupil may be calculated from the magnification of this optical system, namely

$$\begin{aligned} m &= L/L' \\ &= (-371.11)/(-328.06) \\ &= +1.131\,2 \end{aligned}$$

So, if a real pupil has a diameter of 4 mm, it appears to have a diameter of 4(+1.131 2) = 4.52 mm which represents the diameter of the entrance pupil. This, too, is the diameter of the pupil measured by the practitioner.

### 20.3.3 THE EXIT PUPIL

This is defined as the image of the aperture stop formed by that part of the optical system on the eye side. This may be adapted to read: the exit pupil is the image of the real pupil formed by rays refracted by the crystalline lens. Since the exit pupil is also defined as the image of the entrance pupil formed by the complete optical system, it will be clear that only rays entering the system through the entrance pupil will leave the system through the exit pupil. In terms of visual optics we may say that only rays entering the eye through the entrance pupil (and, therefore, passing through the exit pupil) contribute to the formation of the retinal image.

Since the eye may be described as a 'self-contained' system, little practical use is made of the exit pupil. However, to calculate its position and dimension may help to increase the reader's understanding of the optics of the eye.

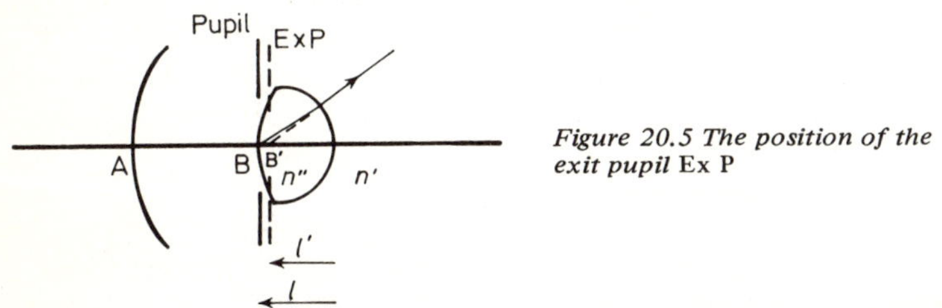


*Figure 20.5 The position of the exit pupil* Ex P

This calculation will be based on Gullstrand's Simplified Schematic Eye No.2 (*see Table 4.1* and *Figure 20.5*). The object distance will be the same as the crystalline lens thickness, thus:

$$L = \frac{n''}{l} = \frac{1413}{-3.6} = -392.50\text{ D} \leftarrow l = -3.6\text{ m}$$

back surface power of lens

$$F = \frac{n' - n''}{r} = \frac{1336 - 1413}{-6.0} = \frac{+12.83}{} +$$

$$L' = -379.67 \rightarrow l' = \frac{n'}{L'} = \frac{1336}{-379.67} = -3.52$$

This means that the exit pupil is found 0.08 mm behind the real pupil.

The diameter of the exit pupil, $h'$, based on a real pupil diameter $h = 4$ mm is given by

$$\begin{aligned} h' &= hm \\ &= h(L/L') \\ &= (4)(-392.50/-379.67) \\ &= (4)(+1.033\,8) \\ &= 4.14 \text{ mm} \end{aligned}$$

As for the second definition of the exit pupil, namely it being the image formed of the entrance pupil by the complete optical system, this can be easily explained. Using the concept of the reversibility of the ray path in an optical system (that is, the image of an object formed by any optical system, may act as object and would then be imaged by that optical system in the plane of the original object, and of the same size), it will be clear that the entrance pupil will be imaged by the cornea in the plane of, and of the same size as the real pupil. The latter, in turn, will be imaged by the crystalline lens to form the exit pupil, as shown above.

## 20.4 Contact lens correction

The calculation of the power (back vertex power) of a contact lens correction is virtually the same as the calculation of the ocular correction when the spectacle correction and vertex distance are known. At

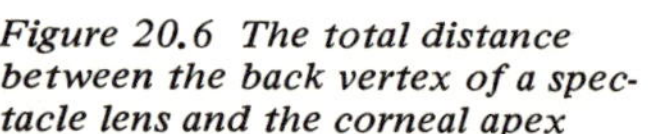

*Figure 20.6 The total distance between the back vertex of a spectacle lens and the corneal apex*

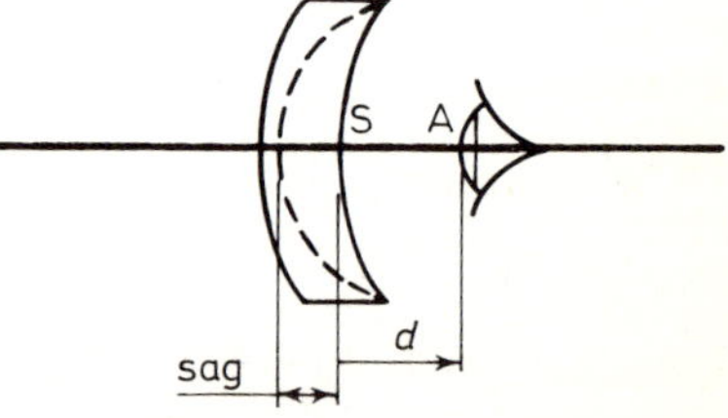


this point the reader is referred to sections 9.2.1 and 9.2.2, and reminded that in the optics of contact lenses, the clinically measured vertex distance should be employed. It is also important to take account of the sagitta (sag, for short) of the spectacle lens since this affects the clinically measured vertex distance (*Figure 20.6*). Consider a spectacle lens

with a −6.00 D back surface power. The sag-values for lenses of the following diameters are:

| diameter (mm) | sag (mm) |
|---|---|
| 40 | 2.33 |
| 50 | 3.67 |
| 60 | 5.33 |
| 70 | 7.34 |

It will be understood that such values become significant with the higher spectacle lens corrections.

The following scheme represents the calculation of the back vertex power of a contact lens correction on the basis of the spectacle lens correction:

| | | | |
|---|---|---|---|
| Spectacle lens power (*BVP*) | = | $F_{sp}$ (D) ⟶ | $f'_{sp}$ (mm) |
| Clinical vertex distance + sag | | | $d$ − |
| Contact lens power (*BVP*) | = | $F_{CL}$ (D) ⟵ | $f'_{CL}$ |

where $f'_{CL}$ = second focal length of the contact lens.

In the course of the fitting of contact lenses a so-called over-refraction is carried out. This is done when an eye has been fitted with a contact lens which produces a correct fit but does not have the required *BVP*. The desired optical correction is then found with the help of a trial frame and trial lenses, or a refractor head or phoropter. The *BVP* of the contact lens that is required, is found by adding the *BVP* of the contact lens used and the power of the trial lens(es). If the latter represents a small correction (say, less than 4 D), the vertex distance of the trial lens will have a negligible effect. However, if the power of the trial lens is more than 4 D, the vertex distance and, where appropriate, the sag of the trial lens, should be taken into account.

## 20.5 Astigmatism

### 20.5.1 INTRODUCTION

As was shown in section 3.3.3, there are several sources of astigmatism in the eye. Details will be discussed in the following sections.

### 20.5.2 CORNEAL ASTIGMATISM

The cornea consists of two refracting surfaces, each of which could be astigmatic. It is generally accepted that when the front surface is astigmatic, the back surface will be astigmatic too. It is very difficult to

obtain details about the latter (*see* sections 2.2.4 and 19.3) and so a certain relationship between front and back surfaces is usually assumed. It is thought that the back surface has a power equal to −1/10th of that of the front surface. *Table 2.1* shows the following relationship for Gullstrand's No.1 eye:

$$\frac{\text{back surface power}}{\text{front surface power}} = \frac{-5.88}{+48.83} = -\frac{10}{83}$$

which is reasonably close to the figure of −1/10th.

The keratometer (also referred to as ophthalmometer: an instrument for the measurement of the radii of curvature of the cornea) was originally designed to measure the total amount of corneal astigmatism since it was thought that corneal astigmatism was almost wholly responsible for the total astigmatism of the eye. Until the advent of the contact lens the radii of curvature of the cornea were of little clinical interest. Since the instrument is based on measuring radii of curvature, a conversion has to be made to obtain the surface powers associated with the radii. The formula for the conversion is equation (1.4):

$$F = (n' - n)/r$$

which applies to a single refracting surface only. Hence, the cornea is assumed to consist of a single refracting surface. The refractive index of the cornea has been assumed to be 1.337 5 (in the majority of instruments; other values used are 1.336 and 1.332). In fact, the refractive index of the cornea is 1.376 (*Table 2.1*). The lower refractive index of 1.337 5 was chosen, amongst other reasons, to take account of the negative power contributed by the back surface of the cornea. The value obtained in this fashion has the following relationship with the actual power of the anterior surface of the cornea:

$$\frac{\text{keratometer power}}{\text{cornea's front surface power}} = \frac{(1.337\,5 - 1)1000/r}{(1.376 - 1)1000/r} = \frac{337.5}{376} = 0.897\,6$$

where $r$ is the anterior corneal radius expressed in millimetres. This shows that the power measured with the keratometer is approximately 9/10th of that of the real power of the anterior surface of the cornea. In this way the negative power of the back surface of the cornea has been taken account of.

The corneal astigmatism calculated from the difference in power measured along the two principal meridians of the cornea, is affected by this discrepancy in the refractive indices, too.

### 20.5.3 LENS ASTIGMATISM

This refers to any astigmatism contributed by the crystalline lens. According to Le Grand (1964) quoting Czellitzer (1927), there appears to be no lens astigmatism in about 50% of the cases, more than 40% shows against-the-rule astigmatism while with-the-rule astigmatism is uncommon. However, when with-the-rule lens astigmatism is present it is usually associated with against-the-rule corneal astigmatism so that the effect may be very small. Lens astigmatism seldom exceeds 2 D (Le Grand, 1964).

### 20.5.4 RESIDUAL ASTIGMATISM

This is defined as the amount of astigmatism that can be measured when a centred, spherical contact lens is placed on the cornea. It should be equal to the difference between corneal astigmatism and spectacle astigmatism (allowing for the vertex distance), but clinical practice shows that this is seldom true (Stone and Phillips, 1972).

### 20.5.5 CORRECTION OF ASTIGMATISM

For various reasons contact lenses with a toroidal back central optic portion and/or a toroidal front central optic portion are fitted from time to time. In addition, some of these lenses may have one or more toroidal peripheral curves. Clinical considerations associated with these lenses will be ignored in the following discussion. We shall deal with three types of toroidal contact lenses, namely those having

1. A spherical front optic and a toroidal back optic portion.
2. A toroidal front optic and a spherical back optic portion.
3. A toroidal front optic and a toroidal back optic portion.

*1. Contact lenses having a spherical front optic and a toroidal back optic portion*

The relationship of the power of the back surface of the liquid lens and the front surface of the cornea, is given by

$$\frac{F_{LL}}{F_{\text{cornea}}} = \frac{(1.336-1)/r_3}{(1.376-1)/r_3} = 0.893\,6 = \sim 0.9$$

Hence, when a spherical contact lens corrects the ametropia in one meridian, only 0.9 of the astigmatism is corrected. However, the remaining 1/10th is corrected by the back surface of the cornea (*see* section 20.5.2). It follows that a contact lens with a toroidal back optic portion fitted on a spherical cornea, creates an astigmatic optical system. This type of astigmatism is referred to as 'induced astigmatism'. It is solely due to the difference in refractive index between the contact lens and the liquid lens. The power of the contact lens–liquid lens surface is given by equation (1.4):

$$F = (n_2 - n_1)/r_2$$

Substituting the appropriate values into this formula, we find:

$$F = (1.336 - 1.49)1000/r_2$$

$$= -154/r_2 \text{ D} \qquad (20.8)$$

where $r_2$ is measured in millimetres.

Since there are two radii of curvature, equation (20.7) has to be applied twice. The difference gives the amount of induced astigmatism. Consider the following hypothetical case.

EXAMPLE 64

A spherical cornea is fitted with a contact lens which has a toroidal back optic surface. The BOR (radii!) are 7.50 and 8.00 mm. Calculate the induced astigmatism.

*Solution 64*

The powers associated with the two radii are each given by equation (20.8), thus:

| BOR 7.50 mm | BOR 8.00 mm |
|---|---|
| $F = -154/(+7.50)$ | $F = -154/(+8.00)$ |
| $= -20.53$ D | $= -19.25$ D |

The induced astigmatism is $-20.53 - (-19.25) = -1.28$ D.

This problem may also be solved on the basis of a modified Rule 1 (*see* section 20.2) which says: for each 0.05 mm difference in BOR, there will be 0.12 D of astigmatism induced (assuming the cornea to be spherical). In this example there is $8.00 - 7.50 = 0.50$ mm difference in BOR. This generates approximately $(0.50/0.05)0.12 = (10)0.12 = 1.2$ D of astigmatism.

Now, if the spectacle correction were spherical, the contact lens correction at the apex plane would be spherical too. However, the contact lens back optic surface is toroidal. This induces astigmatism. It can only be counteracted by a toroidal front optic surface, such that the shorter radius is along the same meridian as the 7.50 mm BOR. In this way, the induced astigmatism will be neutralized. Rotation of such a lens on the cornea will not cause variations in the axis position of the astigmatism since the contact lens–liquid lens system has been made spherical. Westerhout called this a 'compensated bi-toric contact lens' (Stone and Phillips, 1972).

This type of toroidal contact lens may also be fitted on a toroidal cornea. Consider the following problem:

EXAMPLE 65

Corneal radii: 7.50 mm along 180° and 8.00 mm along 90°
BOR: 7.45 mm along 180° and 7.95 mm along 90°
Calculate the astigmatism of this system.

*Solution 65*

*Step 1.* Calculate the corneal astigmatism. The powers along the principal meridians are given by equation (1.4) (or may be found from a conversion table). Using the refractive index of the keratometer, we find:

| *along 180°* | *along 90°* |
|---|---|
| $F = (1.337\,5 - 1)1000/(+7.50)$ | $F = (1.337\,5 - 1)1000/(+8.00)$ |
| $= +45.00$ D | $= +42.19$ D |

which gives $+45.00 - (+42.19) = +2.81$ DC along 180°.

*Step 2.* This is corrected by the back surface of the liquid lens, namely

| | |
|---|---|
| $F = (1 - 1.336)1000/(+7.50)$ | $F = (1 - 1.336)1000/(+8.00)$ |
| $= -44.80$ D | $= -42.00$ D |

which gives $-2.80$ DC along 180°

*Step 3.* The induced astigmatism is given by equation (20.8):

| | |
|---|---|
| $F = -154/(+7.45)$ | $F = -154/(+7.95)$ |
| $= -20.67$ D | $= -19.37$ D |

which gives $-1.30$ DC along 180°. This represents the astigmatism of the system too.

A further complication can be the presence of both residual astigmatism and induced astigmatism along obliquely crossed meridians. The 'quick' way to find the effect of the two obliquely crossed cylinders is, in this case, to carry out a refraction. However, there are two further methods: a graphical solution, known as 'Stokes' construction' (Stokes, 1883), and a mathematical solution. The latter will be found in any textbook on ophthalmic lenses (Bennett, 1968b; Jalie, 1977).

Stokes' construction, as described by Jalie (1977), appears to be the simplest, and is quoted below:

(a) Transpose cylinders, if necessary, so that both are positive in sign.
(b) Choose the cylinder which has the smallest numerical axis direction, for $F_1$. (If one axis is along 180°, call this 0°.)
(c) Construct an origin (*Figure 20.7*) and draw $F_1$ to scale, from the origin along its prescribed axis direction.

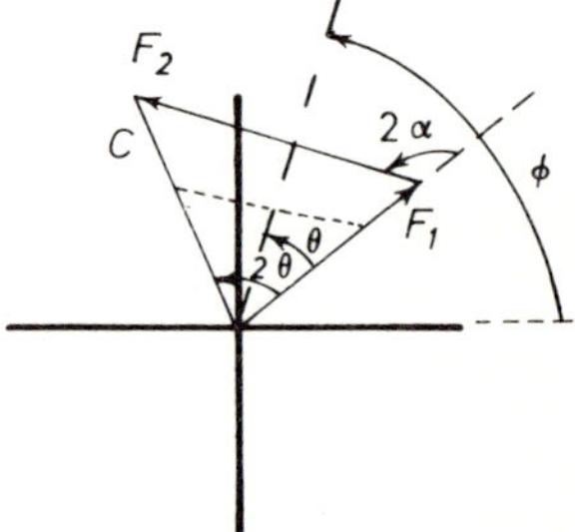


*Figure 20.7 Graphical construction to resolve the powers of obliquely crossed cylinders*

(d) Treating the direction of $F_1$ as the new zero meridian, construct $F_2$ to scale from $F_1$ at the angle 2 (axis $F_2$ − axis $F_1$) = $2a$.
(e) Join $F_2$ to the origin; the length of this line represents the resultant cylindrical component $C$. Since $F_1$ and $F_2$ are positive, $C$ is positive too.
(f) Bisect the angle $2\theta$ between $F_1$ and the resultant cylinder $C$, to obtain the axis direction of the resultant cylinder $C$. This axis direction of $C$ in standard notation is $\phi$ ($= \theta$ + axis $F_1$).
(g) The new spherical component $S$ is given by

$$S = \frac{F_1 + F_2 - C}{2}$$

to which must be added any spherical power derived from the transposition in (a) above.

Returning to example 65, let us assume that in addition to the induced astigmatism of −1.30 DC axis 90°, there is +1.00 DC axis 40° of residual astigmatism. Following Stokes' construction as set out above, we find:

(a) Transposition of the induced astigmatism is required. Thus: −1.30 DS/+1.30 DC axis 180°.
(b) $F_1$ = +1.30 DC axis 180° or +1.30 DC axis 0°.
(c) *See Figure 20.8.*

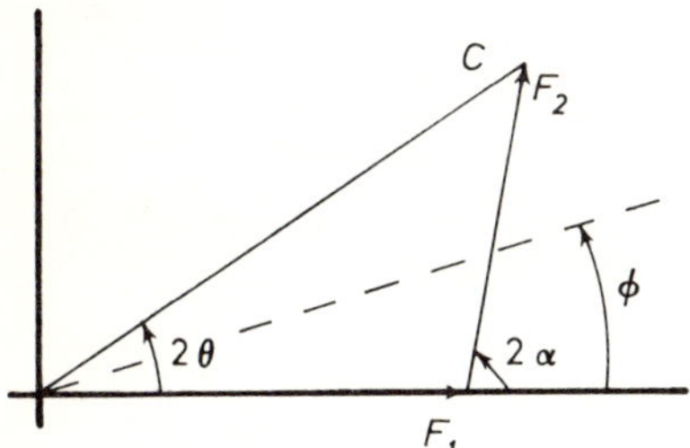


*Figure 20.8 Example 65; scale 5:1*

(d) $2a = 2(40 - 0) = 80°$.
(e) $C$ = +1.8 DC.
(f) $\phi = 15°$.
(g) $S = \dfrac{+1.3 + 1.0 - 1.8}{2} = +0.25$ DS.

Allowing for the spherical power of the induced astigmatism found as a result of the transposition, we find −1.05 DS/+1.8 DC axis 15°. This would be rounded off to −1.00 DS/+1.75 DC axis 15°.

Finally, it is possible to stabilize corneal contact lenses so that rotational movements will not exceed, say, 15°. Scleral lenses are unlikely to be subject to such large rotational movements.

## 2. *Contact lenses having a toroidal front optic and a spherical back optic portion*

The first observation must be that any corneal astigmatism will be neutralized by the liquid lens. Any residual astigmatism can then be corrected by the toroidal front optic portion provided the contact lens does not rotate. The residual astigmatism can best be determined by means of a refraction while a suitable contact lens is worn.

*3. Contact lenses having a toroidal front optic and a toroidal back optic portion*

The need for this type of lens has been indicated in the discussion of contact lenses having a spherical front optic and a toroidal back optic portion, in (1) above.

## 20.6 Magnification

Now we have seen that the entrance pupil is the aperture through which the image forming rays enter the eye, the calculation of the spectacle magnification ought to be reconsidered. The ray directed at the centre of the entrance pupil at B′ (*Figure 20.4*), is the chief ray of the system. Let us now consider its effect on spectacle magnification for a spectacle correction and a contact lens correction.

As has been explained (section 9.2.2) the vertex distance usually employed in visual optics differs from the clinically measured vertex distance in that the latter is measured between the spectacle plane and the corneal apex (*Figure 20.6*). In the following examples we shall calculate separately the spectacle magnification on the basis of the clinically measured vertex distance, the distance SA used in visual optics, and the distance SB′ with respect to the entrance pupil. The spectacle lens correction is +10.00 DS at a clinically measured vertex distance of 10 mm. The sag will be assumed to be as little as 2 mm, so that the distance SA = +12 mm. The calculations using equation (10.7), show the following results.

| *clinical* | *visual optics* | *entrance pupil* |
|---|---|---|
| $SM_R$ = +100.00 mm | $SM_R$ = +100.00 mm | $SM_R$ = +100.00 mm |
| SA = +12.00 − | SP = +13.67 − | SB′ = +15.05 − |
| $AM_R$ = +88.00 | $PM_R$ = +86.33 | $B'M_R$ = +84.95 |
| *SM* = +1.136 4 | *SM* = +1.158 3 | *SM* = +1.177 2 |

approximate differences: visual optics – entrance pupil 2%; clinical – entrance pupil 4%

Similar calculations may be carried out for a contact lens correction. If we take the same example, we can see from the above calculation

that the *BVP* of the contact lens at the corneal apex is $1000/AM_R$ = 1000/+88.00 = +11.36 D. The spectacle magnification values are:

| *visual optics* | | *entrance pupil* | |
|---|---|---|---|
| $AM_R$ | = +88.00 mm | $AM_R$ | = +88.00 mm |
| AP | = +1.67 − | AB′ | = +3.05 − |
| $PM_R$ | = +86.33 | $B'M_R$ | = +85.95 |
| *SM* | = +1.019 3 | *SM* | = +1.023 9 |

approximate difference 0.5%

On comparing the spectacle magnification of the spectacle correction with that of the contact lens correction at the entrance pupil, we find a difference of about 15%. If we make the comparison at the reduced surface (that is, from the column headed 'visual optics'), we find a difference of about 14%. However, in these cases the shape factor has not been taken into account.

*Table 20.1* gives the shape factors of corneal contact lenses calculated from details provided by Madden and Layman (1975) and assumed to have a BOR of +7.80 mm:

TABLE 20.1

| *BVP* (D) | *Centre thickness* (mm) | *Shape factor* |
|---|---|---|
| −1.00 | 0.21 | 1.01 |
| −11.00 | 0.10 | 1.00 |
| +1.00 | 0.25 | 1.01 |
| +8.00 | 0.43 | 1.02 |
| +9.00 | 0.35 | 1.01 |
| +11.00 | 0.38 | 1.01 |

NOTE

Corneal lenses with a *BVP* in excess of −9.00 and of +8.00 D are made with a 'reduced aperture' (that is, the front optic is of lenticular design).

Returning to the earlier discussion of the spectacle magnification and the shape factor, for a +10.00 DS spectacle lens the shape factor is 1.05 (*see* section 14.4, *Table 14.1*) while for the same power corneal

contact lens it is 1.01 (*Table 20.1*). Allowing for these shape factors we now find the following spectacle magnification values:

*Spectacle correction*

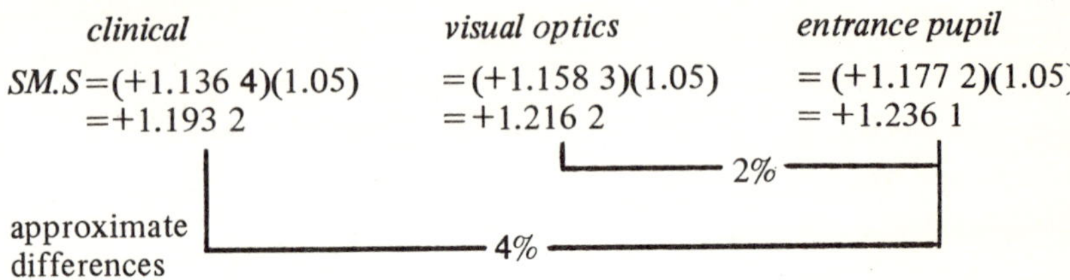


*Contact lens correction*

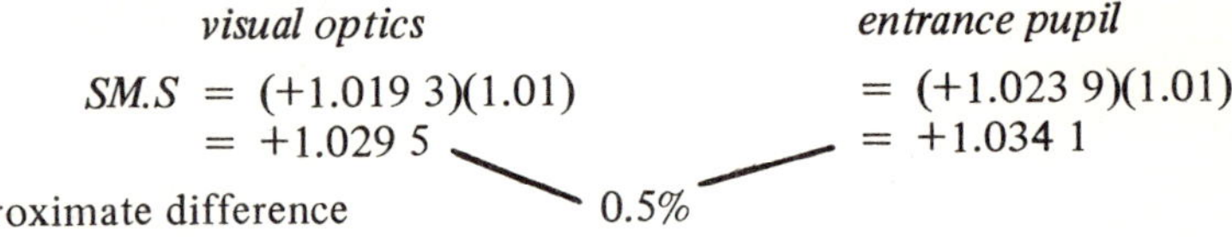


If we now compare the spectacle and contact lens correction spectacle magnification values at the reduced surface and at the entrance pupil, we find differences of approximately 16.5 and 18%. This shows a greater discrepancy than we encountered before. Under certain conditions these small differences may become significant.

Similar calculations may be made for myopic corrections. Since the shape factor is 1.01 (or 1.00) for the contact lens, and 1.01 for the spectacle lens (see *Tables 20.1* and *14.1*), all spectacle magnification values given below would be increased by only 1% (approximately) and may, therefore, be ignored.

The calculations for a −10.00 DS spectacle correction at a clinically measured vertex distance of 10 mm having made allowances for a sag of 4 mm, are as follows:

*Spectacle correction*

| *clinical* | *visual optics* | *entrance pupil* |
|---|---|---|
| $SM_R$ = −100.00 mm | $SM_R$ = −100.00 mm | $SM_R$ = −100.00 mm |
| SA = +12.00 | SP = +13.67 | $SB'$ = +15.05 |
| $AM_R$ = −112.00 | $PM_R$ = −113.67 | $B'M_R$ = −115.05 |
| *SM* = +0.892 9 | *SM* = +0.879 8 | *SM* = +0.869 2 |

approximate differences: 1% (visual optics to entrance pupil); 2% (clinical to entrance pupil)

The *BVP* of a contact lens fitted at the corneal apex needs to be $1000/AM_R = 1000/-112.00 = -8.93$ D. The spectacle magnifications at the reduced surface and at the entrance pupil, will be:

*Contact lens correction*

| *visual optics* | | *entrance pupil* | |
|---|---|---|---|
| $AM_R$ | $= -112.00$ mm | $AM_R$ | $= -112.00$ mm |
| AP | $= +\ 1.67$ − | $AB'$ | $= +\ 3.05$ − |
| $PM_R$ | $= -113.67$ | $B'M_R$ | $= -115.05$ |
| *SM* | $= +0.985\ 3$ | *SM* | $= +0.973\ 5$ |

approximate difference 1%

The difference in spectacle magnification between the spectacle and contact lens correction at the reduced surface and the entrance pupil, is 10.5% in both cases. This shows that this factor is of little importance in myopic corrections.

In general, it should be pointed out that this type of calculation is only worthwhile in cases of very high corrections where significant differences will be found.

NOTE

Comparing the spectacle magnification in the same eye we find that the retinal image size is larger in the myopic and smaller in the hyperopic eye corrected with a contact lens as when corrected with a spectacle lens.

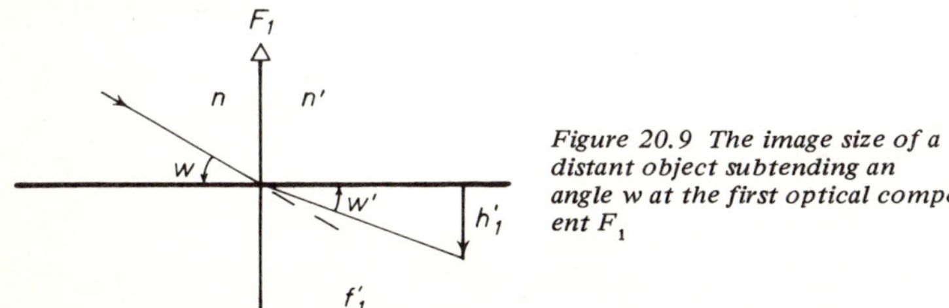


*Figure 20.9 The image size of a distant object subtending an angle w at the first optical component $F_1$*

The above calculations did not take into consideration the magnifying effect of the liquid lens, that is, the shape factor of the liquid lens. A simple way to calculate the shape factor for the contact lens–liquid lens system is the following approach.

From *Figure 20.9* we see that the image size $h'_1$ of a distant object

subtending an angle $w$ at the first optical component $F_1$, is given by

$$\begin{aligned} h_1' &= -f_1' \tan w' \\ &= -\frac{n'}{F_1}\frac{n}{n'}\tan w \\ &= (-n \tan w)/F_1 \end{aligned}$$

If $n = 1$, this equation simplifies to

$$h_1' = (-\tan w)/F_1 \qquad (20.9)$$

Equation (1.13) may be written as

$$h' = mh$$

This may be taken a step further: the size of the $k$th image formed by $k$ – optical components, is given by

$$h_k' = h_1' m_2 \ldots m_k \qquad (20.10)$$

If the equivalent power of such a system is denoted by $F$, equation (20.9) may be altered to

$$h_k' = (-\tan w)/F$$

or

$$F = (-\tan w)/h_k'$$

In view of equation (20.10), this may be written as

$$F = (-\tan w)/(h_1' m_2 \ldots m_k)$$

Having regard to equation (20.9), this latter equation may be written as

$$\begin{aligned} F &= \frac{-\tan w}{[(-\tan w)/F_1]\, m_2 \ldots m_k} \\ &= \frac{F_1}{m_2 \ldots m_k} \end{aligned}$$

Now, the image vergence $L_1'$ of a distant object is given by $L_1' = F_1$. According to equation (1.15),

$$m = \frac{L}{L'}$$

Hence, by substitution, we find

$$F = \frac{L_1'}{(L_2/L_2') \ldots (L_k/L_k')}$$

$$= L_1' \left( \frac{L_2' \ldots L_k'}{L_2 \ldots L_k} \right) \qquad (20.11)$$

And so the shape factor (equation (14.1)) may be written as

$$S = \frac{F_v'}{(L_1'L_2' \ldots L_k')/(L_2 \ldots L_k)}$$

$$= \frac{F_v'(L_2 \ldots L_k)}{L_1' \ldots L_k'} \qquad (20.12)$$

As will be seen from *Figure 20.10*, a contact lens–liquid lens system has three interfaces. The image vergence $L_3'$ emerging from the liquid lens back surface represents the back vertex power of the system.

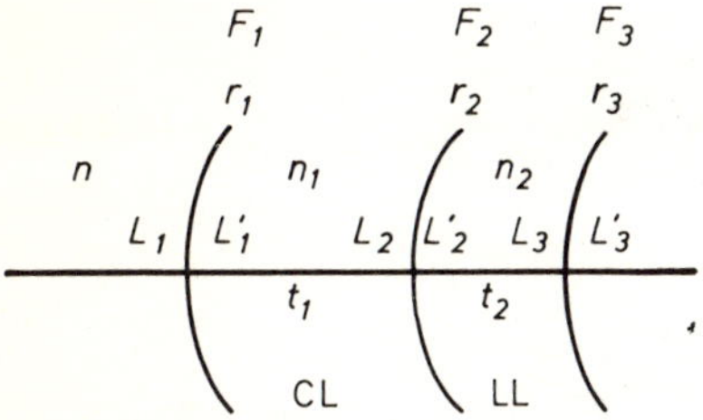

*Figure 20.10 A contact lens–liquid lens system*

Hence, $F_v' = L_3'$. Thus, when equation (20.12) is applied to a contact lens–liquid lens system, we have

$$S = \frac{F_v'L_2L_3}{L_1'L_2'L_3'}$$

$$= \frac{L_3'L_2L_3}{L_1'L_2'L_3'}$$

$$= \frac{L_2L_3}{L_1'L_2'} \qquad (20.13)$$

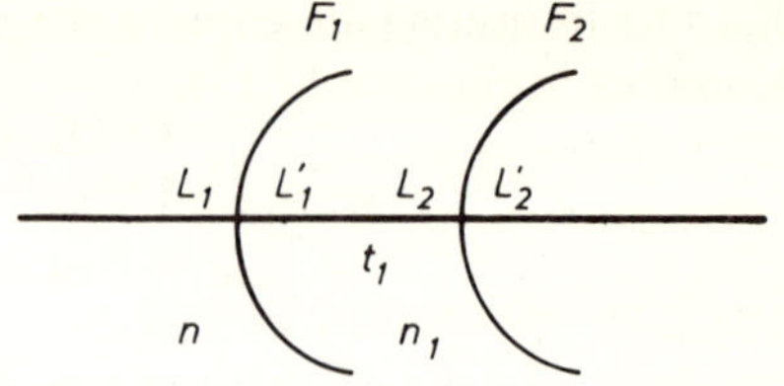


*Figure 20.11 Examples 66, 67, 68 and 69*

The shape factor may be found as set out in the following example (*Figure 20.11*).

EXAMPLE 66

| | |
|---|---|
| *BVP* of scleral contact lens | +8.00 D |
| BOR $= r_2$ | +8.00 mm |
| $t_1$ | 0.80 mm |
| $t_2$ | 0.10 mm |
| 'K'-reading $= r_3$ | 7.85 mm |

Calculate the shape factor of the contact lens–liquid lens system.

*Solution 66*

*Step 1.* Using the step-along method, find the front surface power of the contact lens:

$$BVP = L_2 = +8.00 \text{ D}$$

$$F_2 = \frac{n - n_1}{r_2} = \frac{(1 - 1.49)1000}{+8.00} = -61.25 \quad -$$

$$L_2 = +69.25 \rightarrow l_2 = +14.44 \text{ mm}$$

$$t_1/n_1 = \frac{0.80}{1.49} = +0.54 \quad +$$

$$L_1' = +66.76 \leftarrow l_1' = +14.98$$

$$L_1 = 0.00 \quad -$$

$$F_1 = +66.76$$

(This completes the details of the contact lens.)

*Step 2.* Calculate the vergences $L_2'$ and $L_3$ as follows:

From Step 1,

$$L_2 = +69.25 \text{ D}$$

$$F_2 = \frac{n_2 - n_1}{r_2} = \frac{(1.336 - 1.49)1000}{+8.00} = -18.75 \quad +$$

$$L_2' = +50.50 \rightarrow l_2' = +19.80 \text{ mm}$$

$$t_2/n_2 = \frac{0.1}{1.336} = +\ 0.07 \quad -$$

$$L_3 = +50.68 \leftarrow l_3 = +19.73$$

*Step 3.* Substituting these values into equation (20.13), we obtain

$$S = \frac{(+69.25)(+50.68)}{(+66.76)(+50.50)}$$

$$= +1.041\,0$$

This shows that the shape factor would add about 4% to the retinal image size.

From equation (14.1), the shape factor of the contact lens would be given by

$$S = \frac{BVP}{F_1 + F_2 - (t_1/n_1)F_1F_2}$$

Substituting the value from Step 1 above, we find

$$S = \frac{+8.00}{+66.76 - 61.25 - (0.000\,8/1.49)(+66.76)(-61.25)}$$

$$= \frac{+8.00}{+5.51 + 2.20}$$

$$= +1.037\,6$$

This indicates that the shape factor of the liquid lens adds about 0.3%.

NOTE

The shape factor will vary with the object distance: when the object vergence $L_1 \neq 0$, the image vergence $L_1' \neq F_1$. This will affect all subsequent values. See also the next section.

## 20.7 Afocal lenses

An afocal lens is a lens which forms a distant image of a distant object. In other words: it is of zero power. This does not necessarily mean that the two refracting surfaces are of equal power but of opposite sign, since the thickness of the lens plays a role too, as will be shown in the following example:

EXAMPLE 67

An afocal (scleral) contact lens has a BOR of 8.20 mm and a thickness of 0.70 mm. Calculate the radius of the front surface (*Figure 20.11*).

*Solution 67*

Using the step-along method and working in the direction opposite to that of the incident light, we find:

$$BVP = L_2' = 0.00\text{ D}$$

$$F_2 = \frac{n - n_1}{r_2} = \frac{(1 - 1.49)1000}{+8.20} = -59.76 \quad -$$

$$L_2 = +59.76 \rightarrow l_2 = +16.73\text{ mm}$$

$$t_1/n_1 = \frac{0.70}{1.49} = +0.47 \quad +$$

$$L_1' = +58.14 \leftarrow l_1' = +17.20$$

$$L_1 = 0.00 \quad -$$

$$F_1 = +58.14$$

And so

$$r_1 = \frac{n_1 - n}{F_1} = \frac{(1.49 - 1)1000}{+58.14} = +8.43\text{ mm}$$

This shows that the front radius is flatter than the back radius of this afocal lens.

The shape factor of this lens may be calculated from equation (20.12) which reads in this case:

$$S = \frac{F_v' L_2}{L_1' L_2'}$$

Since $F_v' = BVP = L_2'$ the equation may be simplified to

$$S = \frac{L_2}{L_1'}$$

Substituting these values as found in the above example, we get

$$S = \frac{+59.76}{+58.14}$$

$$= +1.027\,9$$

So far the lens has only been considered for use in distance vision. Under conditions of near vision the effect of the lens is no longer 'afocal' as will be demonstrated in the following example.

EXAMPLE 68

The scleral contact lens of example 67 is used to view an object at a distance of 25 cm. Calculate the vergence emerging from the lens (*Figure 20.11*).

*Solution 68*

Using the step-along method, we find:

$$
\begin{array}{lllcll}
L_1 & = & -4.00\ \text{D} & \longleftarrow & l_1 = & -25.00\ \text{cm} \\
F_1 & = & +58.14 \quad + & & & \\
L_1' & = & +54.14 & \longrightarrow & l_1' = & +18.47\ \text{mm} \\
 & & & & t_1/n_1 = \dfrac{0.70}{1.49} = & +0.47 \quad - \\
L_2 & = & +55.56 & \longleftarrow & l_2 = & +18.00 \\
F_2 & = & -59.76 \quad + & & & \\
L_2' & = & -4.20 & & &
\end{array}
$$

This shows that owing to the afocal lens the vergence reaching the eye has been increased by −0.20 D. This is an unwanted effect as far as presbyopes are concerned; since the negative vergence incident on the afocal lens is enhanced, the eye will have to accommodate more. This may be difficult, if not impossible, when presbyopia is present.

When a positive pencil of light is incident on the same afocal lens we find that the lens has the effect of increasing the (positive) vergence of the pencil too:

EXAMPLE 69

A pencil of +8.00 D vergence is incident on the scleral lens of example 67. Calculate the vergence emerging from the lens (*Figure 20.11*).

*Solution 69*

Using again the step-along method, we obtain:

$$
\begin{array}{lllcll}
L_1 & = & +8.00\ \text{D} & & & \\
F_1 & = & +58.14 \quad + & & & \\
L_1' & = & +66.14 & \longrightarrow & l_1' = & +15.12\ \text{mm} \\
 & & & & t_1/n_1 = & +0.47 \quad - \\
L_2 & = & +68.26 & \longleftarrow & l_2 = & +14.65 \\
F_2 & = & -59.76 \quad + & & & \\
L_2' & = & +8.50 & & &
\end{array}
$$

## 20.8 Relative spectacle magnification

The reader is referred to chapter 11 for the basic principles of relative spectacle magnification. In this section the advantages or disadvantages

of correcting an anisometropic pair of eyes by either spectacle lens(es) and/or contact lens(es), will be discussed.

It is common practice to divide ametropia into either axial or refractive. The same assumption is often made about anisometropia. The reasons for this artificial division have been discussed (section 11.2).

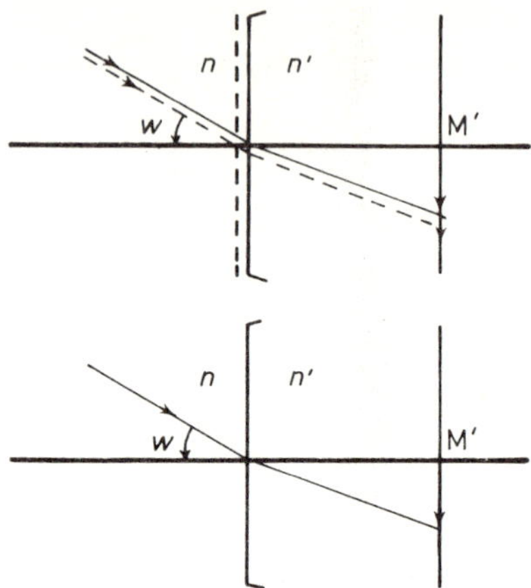


*Figure 20.12 A pair of refractively anisometropic eyes. The vertical interrupted line indicates the position of the contact lens*

Refractive anisometropia will be discussed first. *Figure 20.12* shows a pair of refractively anisometropic eyes. Since their axial length is the same, the basic retinal image sizes are the same too. A contact lens correction will affect the retinal image size only very slightly. A spectacle correction would produce a larger retinal image size difference, since the spectacle magnification (including the shape factor) of a spectacle correction is larger than that of a contact lens. If both eyes are ametropic, both should be corrected by contact lenses.

A particular case of refractive anisometropia is that of unilateral aphakia. In view of what has been said in the previous paragraphs, the aphakic eye should be fitted with a contact lens in order to produce a sharp retinal image as nearly equal to the retinal image size of its phakic fellow-eye. Detailed information, taking into account the pre-aphakic condition, has been given by Bennett (1968a). Only an intra-ocular inplant lens fitted in the aphakic eye, would produce a retinal image size virtually equal to that in the same eye in its phakic condition (Nordlohne, 1975).

In axial anisometropia, the situation is different (*Figure 20.13*). Here the basic retinal images are of unequal size: the longer eye has

the larger basic retinal image. Fortunately, when the powers of the eyes are equal, the longer eye will require a more negative correction. If this eye were corrected by a contact lens, the retinal image size in

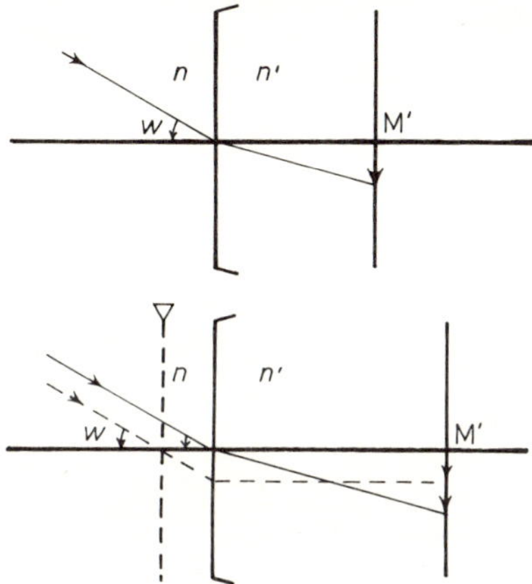


*Figure 20.13 A case of axial anisometropia. The vertical interrupted line indicates the spectacle plane*

that eye would hardly differ from the basic retinal image size (as explained in the paragraphs on refractive anisometropia). It is, therefore, advantageous to fit the eye with the larger retinal image with a spectacle lens. The shorter, fellow-eye, if not emmetropic, should then be fitted with a contact lens to make the retinal images as nearly equal as possible.

The whole object of the establishment of equal retinal image sizes is to facilitate binocular single vision. As a rough guide it may be said that the closer the two retinal image sizes in a pair of eyes approach equality of size, the higher the degree of binocular vision.

## 20.9 Accommodation and presbyopia

As was shown in section 12.9, a myope corrected by a spectacle lens for distance vision will have to accommodate less to see clearly an object placed at a given near vision distance, than an emmetrope. On the other hand, a hyperope corrected by a spectacle lens for distance vision will have to accommodate more than an emmetrope, for the same near vision distance.

When these ametropes are fitted for distance vision with a contact lens correction, it may be equated to making them artificially emmetropic. As a result, myopes will have to accommodate more than with a spectacle correction, and hyperopes need not accommodate as much as with a spectacle correction. In particular myopic near-presbyopes will be affected since their presbyopia will become suddenly more apparent. Prospective contact lens wearers in the pre-presbyopic age group ought to be warned about this problem.

Prospective hyperopic near-presbyopic contact lens wearers will be somewhat better off; when fitted with contact lenses they will notice that they can see clearly a little nearer by than with a spectacle correction. It means also that presbyopia will become apparent a little later when wearing a contact lens correction than when wearing a spectacle correction.

NOTE

These differences in effective accommodation are relatively small and may pass almost unnoticed! This may be verified from the calculations about near vision through afocal lenses, in section 20.7. *See also* Stone (1967) and Stone and Phillips (1972).

The following example shows the calculation sequence for the comparison of the ocular accommodation through a spectacle correction and a contact lens correction.

EXAMPLE 70

An eye is corrected (for distance vision) by a −10.00 DS spectacle lens at 12 mm. Calculate the ocular accommodation required to see

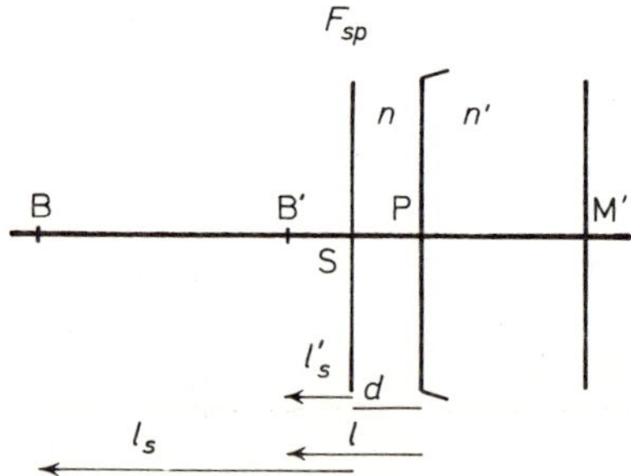


*Figure 20.14 Example 70 showing the distance spectacle correction*

clearly an object situated $\frac{1}{3}$ m in front of the spectacle plane, through the spectacle correction and through the appropriate contact lens correction (assuming the contact lens to be thin) (*Figure 20.14*).

*Solution 70*

*Step 1.* Calculate the ocular correction. This would be the same as the *BVP* of the contact lens.

$$F_{sp} = -10.00\text{ D} \longrightarrow f'_{sp} = -10.00\text{ cm}$$
$$d = +1.2 \quad -$$
$$K = -8.93 \longleftarrow k = -11.20$$

*Step 2.* Calculate the vergence incident on the eye from the near object when the spectacle correction is worn.

$$L_s = -3.00\text{ D} \longleftarrow l_s = -33.33\text{ cm}$$
$$F_{sp} = -10.00 \quad +$$
$$L'_s = -13.00 \longrightarrow l'_s = -7.69$$
$$d = +1.2 \quad -$$
$$L = -11.25 \longleftarrow l = -8.89$$

*Step 3.* Calculate the ocular accommodation. From equation (12.3) we find

$$A = -8.93 - (-11.25)$$
$$= +2.32\text{ D}$$

*Step 4.* Calculate the vergence incident on the eye (that is, the contact lens, assuming the contact lens to be thin). *See Figure 20.15.*

$$l_s = -33.33\text{ cm}$$
$$d = +1.2 \quad -$$
$$L = -2.90\text{ D} \longleftarrow l = -34.53$$

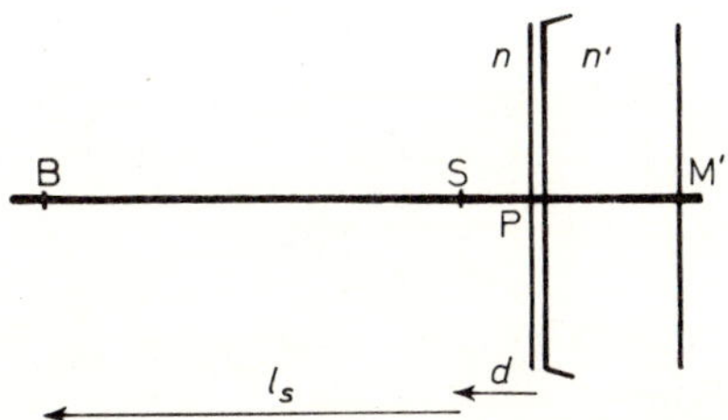


*Figure 20.15 Example 70 showing the scheme for the calculation of the vergence from the near object, incident on the eye*

*Step 5.* The ocular accommodation is now given by equation (12.15), namely $A = -(-2.90) = +2.90$ D.

Thus, ocular accommodation wearing a spectacle correction, is +2.32 D; and ocular accommodation wearing a contact lens correction is +2.90 D.

Now, if we were to assume that the 'K'-reading of the eye were 7.80 mm and that the BOR of the contact lens were of the same radius, the thickness being 0.10 mm, we have the following calculation (*Figure 20.11*):

Calculate the power of the front surface of the contact lens:

$$
\begin{aligned}
BVP = L_2' &= \quad = -8.93\ \text{D} \\
F_2 &= \frac{n - n_1}{r_2} = \frac{(1 - 1.49)1000}{+7.80} = -62.82 \quad - \\
L_2 &= \quad +53.89 \longrightarrow l_2 = +18.56\ \text{mm} \\
&\qquad t_1/n_1 = \frac{0.10}{1.49} = +0.07 \quad + \\
L_1' &= \quad +53.68 \longleftarrow l_1' = +18.63 \\
L_1 &= \quad 0.00 \quad - \\
F_1 &= \quad +53.68
\end{aligned}
$$

Next we can calculate the ocular accommodation through the contact lens, as follows (*Figure 20.16*).

$$
\begin{aligned}
l_s &= -33.33\ \text{cm} \\
d &= -1.2 \\
t_1 &= +0.01 \quad + \\
L = -2.90\ \text{D} \longleftarrow l &= -34.52 \\
F_1 = +53.68 \quad + & \\
L_1' = +50.78 \longrightarrow l_1' &= +19.69\ \text{mm} \\
t_1/n_1 &= \frac{0.10}{1.49} = +0.07 \quad - \\
L_2 = +50.97 \longleftarrow l_2 &= +19.62 \\
F_2 = -62.82 \quad + & \\
L_2' = -11.85 &
\end{aligned}
$$

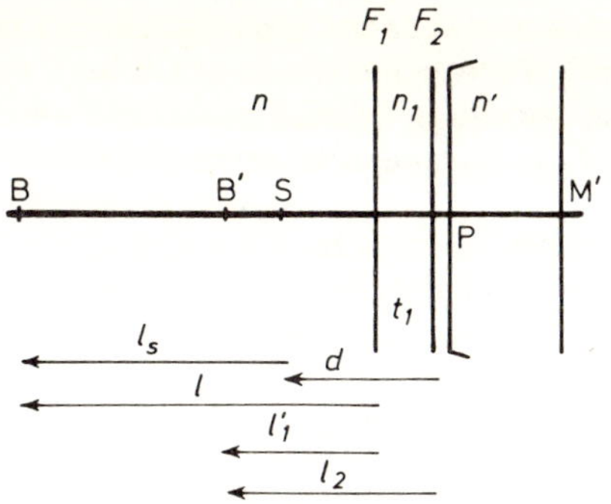


*Figure 20.16 As* Figure 20.15 *where the contact lens is considered to be a thick lens system*

From equation (12.3) where $L = L_2'$, we find that the ocular accommodation is

$$A = -8.93 - (-11.85)$$
$$= +2.92 \text{ D}$$

which is hardly at variance with the simplified calculation shown above.

## 20.10 Binocular vision

This section will deal with differences (if any) in binocular vision encountered when a spectacle wearer becomes a contact lens wearer. The main part of the discussion will revolve around a spectacle correction with lenses centred for distance vision and used for near vision, and contact lenses for distance vision used for near vision.

Spectacle lenses centred for near vision but without a near vision addition, are not likely to be used and need not be discussed. Near vision contact lenses are not likely to be used either; they would require a negative addition (in the form of a spectacle correction) for distance vision.

Let us, therefore, consider a non-presbyopic patient who is wearing a distance vision spectacle correction centred for distance vision, and who now desires to wear contact lenses. As far as distance vision is concerned, there are no optical problems. However, what is happening when the patient wants to do some close work? Is there a change in the relationship between accommodation and convergence when the spectacle correction is replaced by the contact lens correction?

The effect on accommodation has been discussed in the previous section. The effect on convergence has not yet been discussed. For the purpose of this discussion we shall assume that the contact lens will remain centred on the eye for all working distances.

When a pair of myopic eyes corrected by spectacle lenses converges to look at an object situated near by, the visual axes will no longer pass through the optical centres of the lenses and become subject to a prismatic effect, namely base in. Thus, the eyes need not converge as much as when they were not looking through the lenses. At the same time, the eyes need not accommodate as much as when they had been emmetropic. The opposite holds for a hypermetropic pair of eyes corrected by spectacle lenses. On convergence the eyes become subject to prismatic effect, namely base out. This requires more convergence (compared with the amount required from an emmetropic pair of eyes), but corrected hypermetropic eyes must also accommodate more than emmetropic eyes. Thus we find that increased (decreased) convergence goes hand in hand with an increase (decrease) in accommodation. The relationship between convergence and accommodation may be expressed as a ratio. This is usually referred to as the ACA-ratio. Definitions of the ACA-ration vary (compare Davson (1962), Moses (1975) and Schapero *et al.* (1968) with Giles (1960), Ruben (1975) and Stone and Phillips (1972)). For the purposes of this text it is defined as the ratio of accommodation (in dioptres) to the convergence (in prism dioptres) exerted by a pair of eyes for a given distance, or

$$\text{ACA-ratio} = \frac{A\ (\text{D})}{\text{convergence}\ \phi\ (\Delta)} \qquad (20.14)$$

EXAMPLE 71

Calculate the ACA-ratio for a pair of (reduced) eyes corrected by a pair of +10.00 DS spectacle lenses at 12 mm. The centres of projection lie 13 mm behind the reduced surfaces. The object of regard is situated on the midline between the eyes, $\frac{1}{3}$ m in front of the spectacle plane. $PD = 60$ mm.

*Solution 71*

*See Figures 18.6* and *18.7.*

The accommodation may be calculated as follows:

*Step 1.* Calculate the ocular correction:

$$\begin{array}{llllll} F_{sp} & = +10.00\text{ D} & \longrightarrow & f'_{sp} & = & +10.00\text{ cm} \\ & & & d & = & \underline{+1.2\quad} \; - \\ K & = +11.36 & \longleftarrow & k & = & +8.80 \end{array}$$

*Step 2.* The vergence incident on the eye, is:

$$
\begin{array}{llllll}
L_s & = & -3.00\text{ D} & \longleftarrow & l_s & = -33.33\text{ cm} \\
F_{sp} & = & \underline{+10.00} \; + & & & \\
L_s' & = & +7.00 & \longrightarrow & l_s' & = +14.29 \\
 & & & & d & = \underline{+1.2} \; - \\
L & = & +7.64 & \longleftarrow & l & = +13.09
\end{array}
$$

*Step 3.* The ocular accommodation is thus (equation (12.3))

$$
\begin{aligned}
A &= +11.36 - (+7.64) \\
&= +3.72\text{ D}
\end{aligned}
$$

The convergence per eye was calculated in example 55 (section 18.3) to be (nearly) 11Δ. Thus, the convergence of the pair of eyes is 22Δ. The ACA-ratio is (equation (20.14)):

$$
\begin{aligned}
\text{ACA-ratio} &= \frac{+3.72}{22} \\
&= 0.169
\end{aligned}
$$

Following this example, it is assumed that the reader will be able to carry out the calculations for the other conditions. He will then find that in the case of a pair of eyes corrected by a pair of −10.00 DS lenses (other conditions similar to those of example 71), the convergence of the pair of eyes if 13.33Δ and the accommodation is 2.32 D. From equation (20.14) we find that

$$
\begin{aligned}
\text{ACA-ratio} &= \frac{+2.32}{13.33} \\
&= 0.174
\end{aligned}
$$

It will be seen that this result varies but little from that of the example of the pair of +10.00 DS lenses (example 71).

The next step is to see what happens when the patient starts to wear contact lenses. Since we will assume that the contact lenses will remain centred on the eye, no prismatic effect will have to be taken

account of. The calculation of the ACA-ratio will take the following course (*Figure 20.17*).

EXAMPLE 72

A pair of eyes is fitted with contact lenses. Calculate the ACA-ratio for a near vision distance of $\frac{1}{3}$ m from the spectacle plane which is

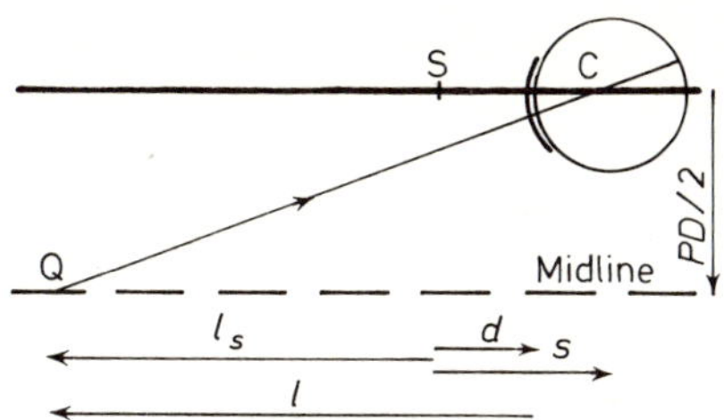


*Figure 20.17 Example 72; calculation of the ACA-ratio*

situated 12 mm in front of the reduced surface. Each centre of projection is at 13 mm behind the reduced surface, and the $PD = 60$ mm.

*Solution 72*

*Step 1.* Calculate the accommodation required. As shown, the effect of a contact lens correction on the vergence incident on the eye is very small (*see* section 20.9). We can, therefore, ignore the effect of the contact lens and simply say:

$$\begin{aligned} A &= -L \\ &= -\frac{n}{l} \\ &= -\frac{1(1000)}{l_s - d} \\ &= -\frac{1000}{-333.33 - (+12)} \\ &= +\frac{1000}{345.33} \\ &= +2.90 \text{ D} \end{aligned}$$

*Step 2.* Calculate the convergence for one eye. For the right eye this is given by (*Figure 20.17*):

$$\tan\phi = -\frac{\tfrac{1}{2}PD}{l-s}$$

$$= -\frac{\tfrac{1}{2}PD}{l_s-d-s}$$

$$= -\frac{-30}{-333-12-13}$$

$$= -\frac{-30}{-358.33}$$

$$= -0.083\,7$$

or $-8.37\Delta$.

*Step 3.* The ACA-ratio is (equation (20.14)):

$$\text{ACA-ratio} = \frac{+2.90}{16.74}$$

$$= 0.173$$

It will be clear that this value applies for all powers of contact lenses used at this near vision distance (all other details being unchanged). It has been shown by Stone (1967) that, for a given near vision distance, the ACA-ratio is virtually the same in a given individual, irrespective of whether a spectacle or contact lens correction is being worn. Bennett (Stone, 1967; Stone and Phillips, 1972) derived an expression which shows that the ratio of

$$\frac{\text{ACA-ratio with spectacle correction}}{\text{ACA-ratio with contact lens correction}} = 1 \text{ (approx.)}$$

Hence, we may conclude that, if a patient did obtain comfortable binocular single vision with a spectacle correction, he should be able to change to a contact lens correction without experiencing any difficulty.

## 20.11 Summary of optical advantages of contact lenses

A contact lens does *not* give rise to:

Astigmatism of oblique pencils.
Distortion.

Chromatic aberration.
A limitation of the field of view.
A blind area in the visual field due to the spectacle frame.
A prismatic imbalance in anisometropia.
Unequal accommodative requirements in anisometropia.
Large spectacle magnification in a corrected aphakic eye.

A contact lens reduced the effect of corneal irregularities (keratoconus, irregular corneal astigmatism, etc.). However, when a contact lens becomes decentred, it produces a prismatic effect (though usually small), and when a toric contact lens rotates, it produces a change in the correction (Visser, 1965; Neumueller, 1968; Raillard, 1969).

# Symbols

The following are the major symbols used throughout this book. The definition of each will be found in the text.

| | |
|---|---|
| $A$ | ocular accommodation |
| $A_s$ | spectacle accommodation |
| $Add$ | addition (near vision) |
| $Amp$ | amplitude of accommodation |
| $B$ | near point vergence |
| $b$ | near point distance |
| $B_s$ | apparent near point vergence |
| $b_s$ | apparent near point distance |
| $d$ | vertex distance |
| $F$ | (equivalent) focal power |
| $f, f'$ | first, second (equivalent) focal length |
| $F_a$ | power of accommodated reduced eye |
| $f_a, f'_a$ | first, second focal length of accommodated reduced eye |
| $F_e$ | reduced eye of non-standard power |
| $f_e, f'_e$ | first, second focal length of reduced eye of non-standard power |
| $F_o$ | reduced eye of standard power |
| $f_o, f'_o$ | first, second focal length of reduced eye of standard power |
| $F_{sp}$ | spectacle lens power |
| $f_{sp}, f'_{sp}$ | first, second focal length of spectacle lens |
| $g$ | pupil diameter |
| $h, h'$ | object, image size |
| $j$ | blur circle diameter |
| $K$ | ocular correction |
| $k$ | far point distance |

| | |
|---|---|
| $K'$ | reduced eye having a non-standard dioptric length |
| $k'$ | reduced eye of non-standard axial length |
| $K'_o$ | reduced eye of standard dioptric length |
| $k'_o$ | reduced eye of standard axial length |
| $L, L'$ | object, image vergence |
| $l, l'$ | object, image distance |
| $L_s, L'_s$ | object, image vergence with respect to the spectacle plane |
| $l_s, l'_s$ | object, image distance with respect to the spectacle plane |
| $M$ | angular magnification |
| $m$ | transverse (lateral) magnification |
| $n, n'$ | object space, image space refractive index |
| $PD$ | interpupillary distance |
| $R$ | curvature |
| $r$ | radius |
| $S$ | shape factor |
| $SM$ | spectacle magnification |
| $RISR$ | retinal image size ratio |
| $RSM$ | relative spectacle magnification |
| $w$ | angle subtended by object at the eye |
| $w'$ | angle subtended by the image inside the eye |
| $w_s$ | angle subtended by object at the spectacle lens |

# Equations

| *Number* | *Equation* | *Page* |
|---|---|---|
| 1.1 | $n \sin i = n' \sin i'$ | 3 |
| 1.2 | $R = \frac{1}{r}$ | 3 |
| 1.3 | $F = (n' - n)R$ | 4 |
| 1.4 | $F = (n' - n)\frac{1}{r}$ | 4 |
| 1.5 | $F = \frac{n'}{f'}$ | 4 |
| 1.6 | $F = \frac{-n}{f} = \frac{n'}{f'}$ | 4 |
| 1.7 | $n'f = -nf'$ | 4 |
| 1.8 | $f = \frac{-n}{F}$ | 4 |
| 1.9 | $f' = \frac{n'}{F}$ | 4 |
| 1.10 | $L = \frac{n}{l}$ | 5 |
| 1.11 | $L' = \frac{n'}{l'}$ | 5 |
| 1.12 | $L' = L + F$ | 6 |
| 1.13 | $m = \frac{h'}{h}$ | 11 |
| 1.14 | $m = \frac{l'}{l}$ | 11 |

| *Number* | *Equation* | *Page* |
|---|---|---|
| 1.15 | $m = \frac{L}{L'}$ | 12 |
| 1.16 | $h' = -f' \tan w$ | 12 |
| 1.17 | $ff' = xx'$ | 13 |
| 1.18 | $F = F_1 + F_2 - dF_1F_2$ | 19 |
| 1.19 | $F = F_1 + F_2 - \frac{d}{n}F_1F_2$ | 19 |
| 1.20 | $F'_v = \frac{n'}{f'_v}$ | 19 |
| 1.21 | $e' = f'_v - f'$ | 20 |
| 1.22 | $e = f_v - f$ | 20 |
| 1.23 | $F'_v = \frac{F}{1-\frac{d}{n}F_1} = \frac{F_1}{1-\frac{d}{n}F_1} + F_2$ | 20 |
| 1.24 | $F_v = \frac{-n}{f_v}$ | 20 |
| 1.25 | $F_v = \frac{F}{1-\frac{d}{n}F_2} = \frac{F_2}{1-\frac{d}{n}F_2} + F_1$ | 20 |
| 1.26 | $\mathrm{N'F'} = \mathrm{FP} = -f$ | 21 |
| 1.27 | $\mathrm{FN} = \mathrm{P'F'} = +f'$ | 21 |
| 1.28 | $\mathrm{NN'} = \mathrm{PP'}$ | 21 |
| 1.29 | $n \sin i = -n \sin i$ | 24 |
| 1.30 | $-\frac{n}{l'} = \frac{n}{l} - \frac{2n}{r}$ | 25 |
| 1.31 | $F = -\frac{2n}{r} = -\left(\frac{n}{l'} + \frac{n}{l}\right)$ | 25 |
| 1.32 | $f' = -\frac{r}{2n}$ | 25 |
| 1.33 | $F = \frac{-n}{f} = \frac{-n}{f'}$ | 25 |

| *Number* | *Equation* | *Page* |
|---|---|---|
| 1.34 | $f' = f$ | 25 |
| 1.35 | $F = -2nR$ | 26 |
| 1.36 | $m = -\frac{l'}{l}$ | 26 |
| 2.1 | $\rho = \left(\frac{n'-n}{n'+n}\right)^2$ | 31 |
| 4.1 | $h' = -k'_o \tan w'$ | 57 |
| 4.2 | $h' = -16.67 \tan w$ mm | 57 |
| 4.3 | $h' = -\frac{3}{4} k' \tan w$ | 57 |
| 5.1 | $K' = \frac{n'}{k'}$ | 58 |
| 5.2 | $K = K' - F_e$ | 59 |
| 5.3 | $K = K'_o - F_o = 0$ | 59 |
| 5.4 | $K = K'_o - F_e$ | 59 |
| 5.5 | $K = K' - F_o$ | 59 |
| 5.6 | $K = K' - F_e$ | 59 |
| 5.7 | $K = \frac{n}{k}$ | 59 |
| 6.1 | $\tan w' = \frac{3}{4} \tan w$ | 63 |
| 6.2 | $m = \frac{3}{4} \times \frac{k'}{k}$ | 64 |
| 6.3 | $m = \frac{K}{K'}$ | 64 |

| *Number* | *Equation* | *Page* |
|---|---|---|
| 8.1 | $j = \left\|\frac{g(l'-k')}{l'}\right\|$ | 75 |
| 8.2 | $j = \left\|\frac{g(K'-L')}{K'}\right\|$ | 76 |
| 8.3 | $j_a = \left\|\frac{4jl}{3k'}\right\|$ | 88 |
| 8.4 | $j_a = \left\|\frac{jK'}{L}\right\|$ | 89 |
| 9.1 | $f'_{sp} = k + d$ | 92 |
| 9.2 | $F_2 = \frac{1}{f'_1 - d}$ | 94 |
| 9.3 | $K = \frac{1}{f'_{sp} - d}$ | 94 |
| 9.4 | $K = \frac{F_{sp}}{1 - dF_{sp}}$ | 95 |
| 9.5 | $F_{sp} = \frac{1}{k + d}$ | 95 |
| 9.6 | $F_{sp} = \frac{K}{1 + dK}$ | 95 |
| 9.7 | $d = \frac{K - F_{sp}}{F_{sp}K}$ | 97 |
| 9.8 | $\delta F_{sp} = F_{sp2} - F_{sp1}$ | 101 |
| 9.9 | $\delta F_{sp} = -\delta d F^2_{sp1}$ | 102 |
| 10.1 | $h'_a = -\frac{3}{4}k' \tan w_o$ | 106 |
| 10.2 | $\tan w = \frac{-h'}{k}$ | 107 |
| 10.3 | $\tan w_o = \frac{-h'}{f'_{sp}}$ | 107 |

| *Number* | *Equation* | *Page* |
|---|---|---|
| 10.4 | $h'' = -\frac{3}{4}k' \tan w$ | 107 |
| 10.5 | $SM = \frac{f'_{sp}}{k}$ | 107 |
| 10.6 | $SM = \frac{K}{F_{sp}}$ | 107 |
| 10.7 | $SM = \frac{\mathrm{SM_R}}{\mathrm{PM_R}}$ | 108 |
| 10.8 | $SM = \frac{1}{1 - dF_{sp}}$ | 108 |
| 10.9 | $SM = 1 + dF_{sp}$ (approx.) | 108 |
| 10.10 | $SM_{\%} = dF_{sp}$ (approx.) | 108 |
| 10.11 | $SM = 1 + dK$ | 109 |
| 10.12 | $h'' = -\frac{3}{4}f' \tan w_o$ | 115 |
| 10.13 | $h'' = \frac{-\tan w_o}{F}$ | 115 |
| 10.14 | $SM = \frac{1 - dL_s}{1 - dL'_s}$ | 120 |
| 11.1 | $RSM_{ax} = \frac{h''_{am}}{h'_{em}}$ | 125 |
| 11.2 | $RSM_{ax} = \frac{F_o}{F}$ | 125 |
| 11.3 | $RSM_{ax} = \frac{F_o}{F_{sp} + F_o - dF_{sp}F_o}$ | 125 |
| 11.4 | $RSM_{ax} = \frac{k'}{k'_o}(SM)$ | 127 |

| *Number* | *Equation* | *Page* |
|---|---|---|
| 11.5 | $RSM_{ax} = \frac{k'}{k'_o} \times \frac{f'_{sp}}{k}$ | 127 |
| 11.6 | $RSM_{ax} = \frac{K'_o}{K'} \times \frac{K}{F_{sp}}$ | 127 |
| 11.7 | $RSM_r = \frac{F_o}{F_{sp} + F_e - dF_{sp}F_e}$ | 127 |
| 11.8 | $RSM_r = SM$ | 127 |
| 11.9 | $RISR = \frac{RSM_R}{RSM_L}$ | 132 |
| 11.10 | $RISR_{ax} = \frac{F_L}{F_R}$ | 135 |
| 11.11 | $RISR_{ax} = \frac{F_{spL} + F_o - d_L F_{spL} F_o}{F_{spR} + F_o - d_R F_{spR} F_o}$ | 135 |
| 11.12 | $RISR_{ax} = \frac{k'_L SM_L}{k'_R SM_R}$ | 135 |
| 11.13 | $RISR_{ax} = \frac{k_L k'_R f'_{spR}}{k_R k'_L f'_{spL}}$ | 135 |
| 11.14 | $RISR_{ax} = \frac{K_R K'_L F_{spL}}{K_L K'_R F_{spR}}$ | 135 |
| 11.15 | $RISR_r = \frac{F_{spL} + F_{eL} - d_L F_{spL} F_{eL}}{F_{spR} + F_{eR} - d_R F_{spR} F_{eR}}$ | 135 |
| 11.16 | $RISR_r = \frac{k_L f'_{spR}}{k_R f'_{spL}}$ | 135 |
| 11.17 | $RISR_r = \frac{K_R F_{spL}}{K_L F_{spR}}$ | 135 |
| 11.18 | $RISR_r = \frac{1 - d_L F_{spL}}{1 - d_R F_{spR}}$ | 135 |

| *Number* | *Equation* | *Page* |
|---|---|---|
| 11.19 | $RISR_r = \frac{1 + d_R K_R}{1 + d_L K_L}$ | 135 |
| 11.20 | $RSM_{ax} = \frac{k'}{k'_o} \times \frac{1 - dL_s}{1 - dL'_s}$ | 137 |
| 12.1 | $F_a = F_e + A$ | 139 |
| 12.2 | $F_a = F_o + A$ | 139 |
| 12.3 | $A = K - L$ | 141 |
| 12.4 | $B = \frac{n}{b}$ | 142 |
| 12.5 | $Amp = K - B$ | 142 |
| 12.6 | $f_a = \frac{-n}{F_a}$ | 144 |
| 12.7 | $f'_a = \frac{n'}{F_a}$ | 144 |
| 12.8 | $r_a = \frac{n' - n}{F_a}$ | 144 |
| 12.9 | $m = \frac{L}{K'}$ | 144 |
| 12.10 | $m = \frac{L}{K'_o}$ | 144 |
| 12.11 | $A_s = -L_s$ | 146 |
| 12.12 | $L = \frac{L_s + F_{sp}}{1 - d(L_s + F_{sp})}$ | 150 |
| 12.13 | $N = 1 + d^2 L F_{sp}$ | 156 |
| 12.14 | $L' = LM^2$ | 159 |
| 12.15 | $A = -L$ | 160 |
| 12.16 | $A = -LM^2$ | 160 |

| *Number* | *Equation* | *Page* |
|---|---|---|
| 12.17 | $A = -L(SM)^2$ | 160 |
| 13.1 | $M = -\frac{l}{f'}$ | 165 |
| 13.2 | $M = \frac{F}{4}$ | 166 |
| 13.3 | $M = \frac{1-dL_s}{1-d(L_s+F)}$ | 169 |
| 13.4 | $M = \frac{F+A_s}{4}$ | 172 |
| 13.5 | $M = \frac{F-L_s}{4}$ | 172 |
| 13.6 | $Add = -(L_s \div \frac{2}{3}Amp)$ | 182 |
| 14.1 | $S = \frac{F'_v}{F}$ | 195 |
| 14.2 | $S = \frac{1}{1-(t/n)F_1}$ | 195 |
| 15.1 | circle of least confusion vergence $\frac{F_1+F_2}{2}$ | 200 |
| 15.2 | circle of least confusion distance $\frac{2n'}{F_1+F_2}$ | 200 |
| 16.1 | $y' = M_y y \quad z' = M_z z$ | 222 |

| *Number* | *Equation* | *Page* |
|---|---|---|
| 16.2 | $y = c \sin \phi \quad z = c \cos \phi$ | 222 |
| 16.3 | $y' = c' \sin \phi' \quad z' = c' \cos \phi'$ | 222 |
| 16.4 | $c' \sin \phi' = M_y c \sin \phi$ | 222 |
| 16.5 | $c' \cos \phi' = M_z c \cos \phi$ | 222 |
| 16.6 | $\tan \phi' = \frac{M_y}{M_z} \tan \phi$ | 222 |
| 16.7 | $c' = c\sqrt{(M_y^2 \sin^2 \phi + M_z^2 \cos^2 \phi)}$ | 223 |
| 16.8 | $c' = cM_y$ | 223 |
| 16.9 | $c' = cM_z$ | 223 |
| 16.10 | $\tan \theta = \frac{\tan \phi \left( \frac{M_y}{M_z} - 1 \right)}{1 + \frac{M_y}{M_z} \tan^2 \phi}$ | 224 |
| 17.1 | $xx' = f_e f_e'$ | 225 |
| 17.2 | $XX' = -(F_e)^2$ | 226 |
| 17.3 | $XX' = -3\ 600.00\ \mathrm{D}$ | 226 |
| 17.4 | $x = k - f_e$ | 226 |
| 17.5 | $x' = k' - f_e'$ | 226 |
| 17.6 | $KX' = -(F_e)^2$ | 226 |
| 17.7 | $XX' = -(F_a)^2$ | 229 |
| 17.8 | $BX' = -(F_a)^2$ | 229 |

| *Number* | *Equation* | *Page* |
|---|---|---|
| 17.9 | $m = 1 - \dfrac{F_e}{K + F_e}$ | 231 |
| 17.10 | $m = 1 - \dfrac{F_a}{K + F_a}$ | 232 |
| 18.1 | $\tan\phi = -\dfrac{\tfrac{1}{2}PD}{l - \mathrm{P_1C}}$ | 234 |
| 18.2 | $\tan\phi = -\dfrac{\mathrm{Q_1Q_2}}{l - \mathrm{P_1C}}$ | 236 |
| 18.3 | $\phi = \dfrac{1}{\mathrm{HQ}}\ \mathrm{MA}$ | 236 |
| 18.4 | $\phi = \dfrac{1}{l - \mathrm{P_1C}}\ \mathrm{MA}$ | 236 |
| 18.5 | $\phi = -\dfrac{\tfrac{1}{2}PD\ (\mathrm{cm})}{\mathrm{HQ\ (m)}}\ \Delta$ | 237 |
| 18.6 | $\phi = -(\text{number of MA})(\tfrac{1}{2}PD \text{ in cm})\Delta$ | 237 |
| 18.7 | $P = cF$ | 238 |
| 20.1 | $F_1 = \dfrac{490}{r_1}$ | 262 |
| 20.2 | $F_2 = \dfrac{-490}{r_2}$ | 262 |
| 20.3 | $F_3 = \dfrac{336}{r_2}$ | 262 |
| 20.4 | $F_4 = \dfrac{-336}{r_3}$ | 262 |
| 20.5 | $F_5 = \dfrac{376}{r_4}$ | 262 |

| *Number* | *Equation* | *Page* |
| --- | --- | --- |
| 20.6 | $L_4' = \dfrac{BVP_{CL} + F_3}{1 - (t_2/n_2)(BVP_{CL} + F_3)}$ | 268 |
| 20.7 | $L_4' = \dfrac{F_3}{1 - (t_2/n_2)F_3}$ | 269 |
| 20.8 | $F = -154/r_2$ | 277 |
| 20.9 | $h_1' = (-\tan w)/F_1$ | 285 |
| 20.10 | $h_k' = h_1' m_2 \ldots m_k$ | 285 |
| 20.11 | $F = L_1' \left( \dfrac{L_2' \ldots L_k'}{L_2 \ldots L_k} \right)$ | 286 |
| 20.12 | $S = \dfrac{F_v'(L_2 \ldots L_k)}{L_1' \ldots L_k'}$ | 286 |
| 20.13 | $S = \dfrac{L_2 L_3}{L_1' L_2'}$ | 286 |
| 20.14 | $\text{ACA-ratio} = \dfrac{A(\mathrm{D})}{\text{convergence } \phi \ (\Delta)}$ | 297 |

# References


AUBERT, H. (1876). *Grundzüge der physiologische Optik.* Leipzig

BANNON, R.E. (1972). 'Near point binocular problems – astigmatism and cyclophoria.' In: *Trans. Int. Ophthalm. Optical Congress 1970.* London: British Optical Ass.

BAUMANN, H.E. (1966). 'Muscular asthenopia and headache in manifest and latent disturbances of binocular vision.' *Zeiss Mitteilungen* **57**, 133–184

BENNETT, A.G. (1956). *Optics of contact lenses.* London: Ass. Dispensing Opticians, 2nd edn

BENNETT, A.G. (1968a). 'The corrected aphakic eye: a study of retinal image sizes.' *Optician*, 106–111, Feb. 2, & 132–135, Feb. 9

BENNETT, A.G. (1968b). *Emsley and Swaine's Ophthalmic Lenses*, vol. 1, London: Hatton Press

BENNETT, A.G. (1972). 'A note on Newton's relation.' *Ophthalm. Optician* **12**, 943–944

BENNETT, A.G. (1976). 'Power changes in soft contact lenses due to bending.' *Ophthalm. Optician* **16**, 939–945

BIER, N. (1956). 'A study of the cornea in relation to contact lens practice.' *Am. J. Optom.* **33**, 291–304. Cited by Stone and Phillips, 1972

BIER, N. (1957). *Contact Lens Routine and Practice.* London: Butterworth

BIER, N. and LOWTHER, G.E. (1977). *Contact Lens Correction.* London: Butterworth

BINKHORST, C.D. and LOONES, L.H. (1975). Resultaten van individuele lensberekeningen. *Ned. T. Geneesk.* **119**, 2069

BLIX, M. (1880). 'Oftalmometriska studier.' *Upsala Läkareförerings Förhandlingar* **15**, 349. Cited by Le Grand, 1956

BS 3521 (1962). *Glossary of terms relating to ophthalmic lenses and spectacle frames.* London: British Standards Institute

CAMPBELL, F.W. (1957). 'The depth of field of the human eye.' *Optica acta* **4**, 157–164

CAMPBELL, F.W. (1959). 'The accommodative response of the human eye.' *Brit. J. physiol. Optics* **16**, 188–203

CHASTON, J.M. (1975). Private communication

COATES, W.R. (1955). 'Amplitudes of accommodation in South Africa.' *Brit. J. physiol. Optics* **12**, 76–81 and 86

COGAN, D.G. (1951). 'Applied anatomy and physiology of the cornea.' *Trans. Am. Acad. Ophthalm.* **55**, 329–359

CZELLITZER. (1927). *Klin. Mbl. Augenheilk.* **79**, 301. Cited by Le Grand, 1964

DAVSON, H. (1962). *The eye*, vol. 3. New York/London: Academic Press

DONDERS, F.C. (1864). *On the anomalies of accommodation and refraction of the eye.* London: New Sydenham Soc. Reprinted: London: Hatton Press, 1952

DREW, N.J. (1941). 'Norms of refraction. A partial historical survey of reported data.' *Am. J. Optom.* **18**, 97–111. Cited by Weale, 1963

DUANE, A. (1912). 'Normal values of the accommodation at all ages.' *J. Am. med. Ass.* **59**, 1010–1013

DUKE-ELDER, S. (1961). *System of ophthalmology*, vol. II. London: Kimpton

DUKE-ELDER, S. (1970). *System of ophthalmology*, vol. V. London: Kimpton

EMSLEY, H.H. (1955). *Visual optics*, vol. 1. London: Hatton Press. 5th edn

EMSLEY, H.H. (1956). *Aberrations of thin lenses.* London: Constable

FINCHAM, E.F. (1935). 'A study of accommodation by photography of the living lens and ciliary body in a case of aniridia.' *Trans. ophthal. Soc. UK* **55**, 145–158

FINCHAM, E.F. (1937). 'The mechanism of accommodation.' *Brit. J. Ophthal.*, monograph no. 8,

FINCHAM, W.H.A. and FREEMAN, M.H. (1974). *Optics.* London: Butterworth. 8th edn

FLETCHER, R.J. (1952). 'Astigmatic accommodation.' *Brit. J. physiol. Optics* **9**, 8–32

FORD, M.W. (1976). 'Computation of the back vertex powers of hydrophilic lenses.' In: *Interdisciplinary Conf. Contact Lenses*, The City University, London. June 24 and 25. 52–54

FREYTAG, G. (1907). *Vergleichende Untersuchungen über die Brechungsindizes der Linse und der flüssigen Augenmedien des Menschen und höherer Tiere in verschiedenen Lebensaltern.* Wiesbaden (Thesis). Cited by Schober, 1970

FRY, G.A. (1940). 'Significance of fused cross cylinder test.' *Optom. Weekly*, Feb. 15

FRY, G.A. and HILL, W.W. (1962). 'The center of rotation of the eye.' *Am J. Optom.* **39**, 581–595

GAUSS, J.K.F. (1838–1843). 'Dioptrische Untersuchungen.' *Abh. kön. Ges. Wiss. Göttingen* **1**. Also: Sonder Abdruck, Göttingen, 1841. Cited by Donders, 1864

GILES, G.H. (1960). *The principles and practice of refraction.* London: Hammond, Hammond & Co.

GRAVES, P.-M. (1968). 'Cours d'optique physiologique et d'optometrie.' Fascicule I: *L'oeil et ses defauts optique.* Paris: Riber

GULLEY, D.E. (1975). Personal communication

GULLSTRAND, A. In: Helmholtz, 1909–1911

HALASS, S. (1959). 'Aniseikonic lenses of improved design and their application.' *Austr. J. Optom.* **42**, 387–393

HELMHOLTZ, H. von. (1858). 'Physiologische Optik.' In: *Karsten's Allgemeine Encyclopaedie der Physik.* Leipzig

HELMHOLTZ, H. von. (1909–1911). *Physiologische Optik.* Hamburg/Leipzig: Voss. 3rd edn. See: Southall, 1924

HUGGERT (1946). Cited by Le Grand, 1964

HUYGENS, C. (1653). *Opuscula posthuma, Tomus primus: Dioptrica.* Reprinted: *Oeuvres complètes de Christiaan Huygens.* La Haye: M. Nijhoff, 1916. Earlier edition: Amsterdam: Janssonio - Waesbergios, 1728

IVANOFF, A. (1953). *Les aberrations de l'oeil.* Paris: Revue d'Optique

JALIE, M. (1977). *The principles of ophthalmic lenses.* London: Ass. Dispensing Opticians. 3rd edn

KAUFMAN, L. (1974). *Sight and mind.* New York/London/Toronto: Oxford University Press

KNAPP, H. (1870). 'The influence of spectacles on the optical constants and visual acuteness of the eye.' *Arch. Ophthalm. Otolaryngol.* **1**, 2. Cited by Michaels, 1975

LE GRAND, Y. (1956). *Optique physiologique*, vol. III. Paris: Revue d'Optique

LE GRAND, Y. (1962). 'Studies on the human cornea: application to the aphakic eye.' *Trans. Int. Ophthalm. Optical Congress 1961.* London: Crosby Lockwood

LE GRAND, Y. (1964). *Optique physiologique*, vol. I. Paris: Revue d'Optique. 3rd edn

LEINHOS, R. (1959). 'Die Altersäbhangigkeit des Augenpupillendurchmessers.' *Optik* **16**, 669–671. Cited by Weale, 1963

LISTING, J.B. (1853). 'Zur Dioptrik des Auges.' In: *Wagner's*

*Handwörterbuch der Physiologie* **4**, 451–504. Braunschweig: F. Bieweg u. Sohn

MADDEN, P. and LAYMAN, R. (1975). *Technical Bulletin.* St. Leonards-on-Sea (Sussex): Madden Contact Lenses

MATTHIESSEN, L. (1877). *Grundriss der Dioptrik geschichteter Linsensysteme.* Leipzig

MATTHIESSEN, L. (1891). *Die neueren Fortschritte in unserer Kenntnis von dem optischen Baue des Auges der Wirbeltiere.* Hamburg

MAUTHNER, L. (1876). *Vorlesungen über die optischen Fehler des Auges.* Wien

MICHAELS, D.D. (1975). *Visual optics and refraction.* St. Louis: Mosby

MOSER, L. (1844). 'Uber das Auge.' In: *Dove's Repertorium der Physik* **5**, 289. Cited by Le Grand, 1964

MOSES, R.A. (Ed.). (1975). *Adler's Physiology of the eye.* St. Louis: Mosby, 6th edn

MOSS, S.G. 'Observations of retinal images in a model eye.' *Optician*, 4–9, June 11 1976 and 8–11, Jan. 21 1977

NEUMUELLER, J.F. (1968). 'The optics of contact lenses.' *Am. J. Optom.* **45**, 785–796

NORDLOHNE, M.E. (1975). *The intraocular implant lens.* The Hague: Dr W. Junk

NORN, M.S. (1974). *External eye: Methods of examination.* Copenhagen: Scriptor

OBSTFELD, H. (1964). 'Notes on near vision astigmatism.' *Optica int.* **1**, 43–46

OBSTFELD, H. (1972). 'Newton' equation in visual optics.' *Ophthalm. Optician* **12**, 713–715

OBSTFELD, H. (1975). 'Three visual optics nomograms.' *Optician*, 4–8 Nov. 21

OBSTFELD, H. (1976a). 'Nieuwe ideeën omtrent de werking van de kruiscylindermethode.' *Oculus, Amsterdam*, 54–58, Mar. 'Das Kreuzzylinder-Verfarhren neu überprüft. *Optometrie* **24**, 4–12

OBSTFELD, H. (1976b). 'The corrected ametropic eye as a telescopic system – 1 & 2.' *Optician*, 4–5, May 21 & 13–16, May 28

OGLE, K.N. (1964). *Researches in binocular vision.* New York/ London: Hafner

PIPE, D.M. (1975). Private communication

PIRENNE, M.H. (1967). *Vision and the eye.* London: Chapman & Hall. 2nd edn

PURKINJE, J.E. (1823). *Commentatio de examine physiologico organi visus et systematic cutanie.* Breslau. Cited by Duke-Elder, 1970

RAILLARD, G. (1969). *La réfraction dans l'optique de contact.* Paris: Livroptic

RAMBO, V.C. (1953). 'Further notes on the varying ages at which different people develop presbyopia.' *Am. J. Ophthalm.* **36**, 709–710. Cited by Weale, 1963

RAMBO, V. (1960). *Proc. All-Idia ophthalm. Soc.* **17**, 263. Cited by Duke-Elder, 1970

RANTZSCH, H. (1964). 'Die Schärfetiefe des menschlichen Auges und ihre Auswirkung auf die Korrektion.' *Augenoptik* 3 & 4. Cited by Trotter, 1967

RAPHAEL, J. (1961). 'Accommodation variations in Israel.' *Brit. J. physiol. Optics* **18**, 181–185

REINER, J. (1965a). 'Zur Theorie der Korrektion achsensymmetrischer und astigmatischer Augen.' *Klin. Mbl. Augenheilk.* **147**, 564–574

REINER, J. (1965b). 'Anamorphotische Abbildungen beim korrigierten astigmatischen Augen.' *Optik* **23**, 26–35

REINER, J. (1966). 'Das korrigierte astigmatische Auge als anamorphotisches System.' *Klin. Mbl. Augenheilk.* **148**, 898–906

REINER, J. (1972). *Auge und Brille – Beiträge zur Optik des Auges und der Brille.* Bucherei des Augenarztes, vol. 59. Stuttgart: Enke

RUBEN, M. (1975). *Contact lens practice.* London: Baillière Tindall

RUBIN, M.L. (1967). 'Surgical procedures available for influencing refractive error.' In: *Refractive anomalies of the eye.* NINDB monograph no. 5. Bethesda (Md): US Department of Health, Education and Welfare

SANSON, L.J. (1837). *Leçons sur les maladies des yeux.* Paris. Cited by Le Grand, 1956

SCHAPERO, M., CLINE, D. and HOFSTETTER, H.W. (Eds). (1968). *Dictionary of visual science.* Radnor (Penn.): Chilton. 2nd edn

SCHOBER, H. (1970). *Das Sehen,* Band I. Leipzig: VEB Fachbuchverlag. 4th edn

SIEBECK, R. (1960). *Optik des menschlichen Auges.* Berlin/Göttingen/Heidelberg: Springer-Verlag

SORSBY, A., BENJAMIN, B., DAVEY, J.B., SHERIDAN, M. and TANNER, J.M. (1957). 'Emmetropia and its aberrations.' *Spec. Rep. Ser. med. Res. Coun. Lond.,* No. 293

SORSBY, A., LEARY, G.A. and RICHARDS, M.J. (1962). 'The optical components in anisometropia.' *Vision Res.* **3**, 43–51

SOUTHALL, J.P.C. (1924). *Helmholtz's Treatise on physiological optics.* Reprinted: New York: Dover, 1962

STEIGER, A. (1895). *Beiträge zur Physiologie und Pathologie der Hornhautrefraktion.* Wiesbaden: J.-F. Bergmann. Cited by Le Grand, 1956

STOKES, G.G. (1883). *Mathematical and Physical Papers* **2**, 172–175. Cambridge: Cambridge University Press. Cited by Bennett, 1968b

STONE, J. (1967). 'Near vision difficulties in non-presbyopic corneal lens wearers.' *The Contact Lens* **1**, 14–16, 24–25

STONE, J. and PHILLIPS, A.J. (1972). *Contact lenses. A textbook for practitioner and student.* London: Barrie & Jenkins

SWAINE, W. (1956). 'Optics of contact lenses and their prescription.' *Brit. J. physiol. Optics* **13**, 149–163

THOMAS, C.I. (1955). *The cornea.* Springfield (Ill.): Thomas

TREUTLER. (1902). 'Einige Bemerkungen zu den schematischen Augen.' *Klin. Mbl. Augenheilk.* **40**, 1 . Cited by Gullstrand, 1909–1911

TROTTER, J. (1967). *Das Auge.* Solothurn: Optik-Verlag. 2nd edn

TSCHERNING, M. (1898). *Optique physiologique.* Paris. Cited by Le Grand, 1964

VISSER, G.J. (1965). *Medische en technische aspecten van de toepassing van contactlenzen.* Oosterbeek (Holland): Adremo

WEALE, R.A. (1963). *The aging eye.* London: Lewis & Co.

WESTERHOUT, D. (1970). 'A clinical survey of corneal oedema, its signs, symptoms and incidence.' *Trans. Contact Lens Soc.,* Presidential Address, Sept. 21

# Index

AF615364